Processing, Preservation and Product Development Techniques for Fruits and Vegetables

NEW INDIA PUBLISHING AGENCY
New Delhi-110 034

About the Author

Prof. Sewa Lal Bhardwaj was born on 1st Feb. 1946 in a small farmer's family of the small village name Darra, Tehsil-Chunar (earlier known as Chunar Garh) of District Mirzaprur, Uttar Pradesh, India. His schooling was from Madhav Vidya Mandir, Purusottampur upto 8th classes. Then he was admitted in 9th class in Central Hindu School and then in Central Hindu College, Kamakshya, Banaras Hindu University, Varanasi. Uttar Pradesh, India. After completing his education in Central Hindu College, he was admitted in B.Sc. (Ag.) and then M.Sc. (Ag.) in Hort and Ph.D. (Hort) in the Department of Horticulture, Faculty of Agriculture, Banaras Hindu University, Varanasi, Uttar Pradesh, India. He completed his M.Sc. and Ph.D. in Horticulture from the department. During study he was a recipient of I.C.A.R. Scholarship in B.Sc. (Ag.), National Loan Scholarship in M.Sc. (Ag.), Research Fellowship from the Institute of Ayurvedic Medicine, Institute of Medical Sciences, Banaras Hindu University, Research Fellowship of C.S.I.R. (Council of Scientific and Industrial Research) and Research Fellowship of U.G.C. (University Grant Commission), New Delhi during Ph.D. Degree. After completing Ph.D. Degree in the year 1974, he joined as a Post-Doctorate Fellow under the Auspices of University Grant Commission, New Delhi, India in the Department of Horticulture, Faculty of Agriculture, Banaras Hindu University. In June 1975 he joined as Assistant Vegetable Botanist (Vegetable Seed Production) in the Department of Vegetable Science, College of Agriculture, C.C.S. Haryana Agricultural University, Hisar (Haryana), India. He has worked on various positions as Assistant Vegetable Botanist (Seed Production), Scientist and Sr. Scientist/Professor till January, 2006. In the mean time he was offered foreign assignment in Botswana from March 1988 to Feb. 1992 under Unified Teaching Service (UTS), Ministry of Education, Government of Botswana, Gaborone, Botswana. After retirement in Feb. 2006, he joined as a Professor of Horticulture, Department of Plant Science, College of Agriculture, Haramaya University, Ethiopia in October 2006. Again he got offer and joined as a Professor of Horticulture, Department of Horticulture and Plant Science, College of Agriculture and Veterinary Medicine, Jimma University, Jimma, Ethiopia in November 2009 and completed my assignment in July 2013. After completing this assignment in Ethiopia, he was offered to join as a Professor of Pomology in the Department of Horticulture, Hamelmalo Agricultural College, Hamelmalo, Keren, Eritrea in March 2014 and completed this assignment in July 2018. At present he is living with his children at Hisar, Haryana, India. Having a long journey in the field of Horticulture and research and teaching experience of different countries. He has planned to present his experiences in manuscript forms. This manuscript is one of them.

Processing, Preservation and Product Development Techniques for Fruits and Vegetables

Sewa Lal Bhardwaj
M. Sc. (Ag.) & PhD. in Horticulture
Retired Sr. Scientist/Professor
C.S.S. Haryana Agricultural University
Hisar-125 005, Haryana, India

NEW INDIA PUBLISHING AGENCY
New Delhi-110 034

nipa®

NEW INDIA PUBLISHING AGENCY
101, Vikas Surya Plaza, CU Block, LSC Market
Pitam Pura, New Delhi – 110 034, India
Email: info@nipabooks.com
Web: www.nipabooks.com

For customer assistance, please contact
Phone: + 91-11-27 34 17 17 Fax: + 91-11- 27 34 16 16
E-Mail: feedbacks@nipabooks.com

ISBN: 978-93-90175-07-9

Composed and Designed by NIPA.

Dedicated to

My wife Late Mrs. Sundari Bhardwaj

Preface

In the modern wave of economic development, horticulture has become a major part and plays very important role in development of agriculture enterprise in world economic development. Among horticultural crops, fruits and vegetables are major crops and considered presently as important food commodities and economical crops. Fruits and vegetables have also important place in our daily diet as they are richest and cheapest source of all essential nutrients such as carbohydrates, fat, protein, vitamins and minerals required by our body for proper functioning of body organs and maintaining good health. Besides these nutrients they also supply water and fibres to our body which play vital roles in our body. FAO WHO (2003) has recommended 400 g/capita/day (100 g fruits and 300 g vegetables) of fruit and vegetables to be consumed for good health and prevent chronic diseases.

Fruits and vegetables become glut in harvesting season due to high productivity and all cannot be consumed at that time. They cannot be stored because there is scarcity of storage facilities which results high wastage of their highly perishables nature. On the other hand there is scarcity of fruits and vegetables in off season. If they are available, the market prize is very high and people can afford it and mostly common people are deprived of consuming fruits and vegetables. Therefore, it is very essential to create facilities and preventive measures to reduce wastage and make more availability of fruits and vegetables to consumers. Other care must be taken after harvesting to prevent initial losses, because losses start immediately after harvesting. To reduce the losses and to make availability of fresh fruits and vegetable during off season, the best mean is the processing and preservation by using simple techniques. There are many processing methods that can be used by small-scale handlers, including drying, fermenting, canning, freezing, preserving and juicing.

Fruit and vegetable processing and preservation has its own place and introduced as a core course of Horticulture Science since long time probably before 1960 s in B. Sc. (Ag.) and M. Sc. (Ag.) Horticulture. But in modern society people have realized the nutritional importance of fruits and vegetables and their processed products and at present it has also been included in Food Science and Technology as a course. Not only inclusions of as a course but several training centers have been established in the country. Keeping these views, the informations have been compiled to provide adequate knowledge about processing and preservation of fruits and vegetables to students, teachers, traders and trainees. With these objectives the manuscript has been prepared in brief and simple language so that every user can understand easily. It will also be very useful to Housewives who create more taste and flavor to our meal.

On the basis of my working and teaching experience, manuscript is not only be useful to Indian community but also very valuable informations for African countries like Ethiopia and Eritrea where people do not known about these technology. Processing and preservation of fruits and vegetables is a new course in their Horticulture development which will help in many ways in the food security and malnutrition.

Author

Aknowledgement

The idea of compiling this manuscript in the form of a book has been developed when author was teaching this course under post-harvest technology of fruits and vegetables during working period in India as well in Ethiopia and Eritrea. Processing and product development of fruits and vegetables as a course has been included in the Syllabus of Horticultural Science of Hamelmalo Agricultural College. While teaching in Eritrea, I was encouraged by teachers and Head of Department of Horticulture to compile theses informations. I am very much thankful to Dr Brahn Khair Sahel director of Labs and a senior teacher of the Horticulture of Hamelmalo Agricultural College, Eritrea., who was always encouraging and advising me to bring as a well known document what you have experience about this field of horticulture. Author is also thankful to Mr. Belay (H.O.D.) for his encouragement and very positive remarks on this course during practical classes. I must not forget to thank instructor Ms Genet my colleague who was attached with me in practical class for her very positive cooperation and arrangement of materials and taking photographs.

I also acknowledge the authors of books and literatures related to the fruit and vegetable preservation and development products.

I am also very thankful to New India Publishing Agency, Pitam Pura, New Delhi, India for publication of this useful manuscript.

Lastly, I am very thankful to my family member for providing helps and facilitating during compilation this manuscript.

Contents

1

Introduction

Agriculture has been the base of human livelihood since ancient time when the man started planting of seeds of some edible plants now called Agricultural crops. Domestication of food plants and animals was started with the development of human civilization. Thus agriculture can be explained as science and technology of growing food crops and rearing of animals. Still in several countries of the world, most of population is based on agriculture for their livelihood particularly in rural areas of that country. According to Williams and Haq (2002) there are over 7000 species of plants that can be used for food. The edible plants species grown for food are called agricultural crops. The crops grown in garden are known as garden cops or Horticulture crops. Horticulture is a branch of agriculture, which deals with intensive cultivation of garden crops such as fruits, vegetables, flowers, ornamental plants, plantation crops, spices, medicinal and aromatic plants, and mushroom. It also deals with post-harvest technology of horticultural crops. Processing and Preservation of fruits and vegetables is a part of Post-harvest Technology of horticultural crops which deals with methods of processing and preservation of these crops to extend the shelf life of fruits and vegetables after harvesting.

All horticultural crops have their own importance in our society. Fruits, vegetables and spices are major ingredients of our meal. Spices add taste and flavor to our food and also have high medicinal values. Flowers and ornamental plants are used for decoration of environment and landscape. Aromatic and plantation crops have become industrial crops. Mushroom growing is a great enterprise. Among these crops fruits and vegetables are major crops and have been considered presently as highly economical and important food crops. There are 1200 crop species are used for vegetable purpose in the world (Hazra and Som, 2006). Fruits and vegetables are richest and cheapest source of all essential nutrients (carbohydrates, fat, protein, vitamins and minerals) required by our body for proper functioning of body organs and maintaining good health. Besides these nutrients they also supply water and fibers to our body which play vital roles in our body. Fruits and vegetables have important place in our daily diet and based on survey made by experts FAO WHO (2003) has recommended 400 g/capita/day of fruit and vegetables (fruits -100 g/capita/day and vegetables -300 g/capita/day) to be consumed for good health and prevent chronic diseases. Therefore, the demand of these food commodities has increased in the modern society due to increased awareness of the nutritional importance of fruits and vegetables. So that people must be aware that what part of fruit and vegetable is to be consumed? (Table 1 and 2). Botanically fruit is a mature ovary of flower formed after pollination and fertilization. Whereas vegetable may be any part of edible plant i.e. root, stem, leaf, flower, flower buds, fruit and seed used for culinary purpose. Fruits are mostly eaten as raw whereas most of vegetables are consumed as raw as salad or cooked forms.

Different Edible Parts of Fruits and Vegetables

There are various parts of fruits and vegetables are consumed by us so that we must know that part to avoid loss of that edible portion of fruits and vegetables. Because many times when we remove unwanted parts or skin while peeling a lot of valuable edible parts are removed.

Table 1: Common and scientific names and edible parts of some common fruits

Name of fruit	Scientific name	Edible parts
Apple	*Pyrus malaus*	Fleshy thalamus
Banana	*Musa paradisica var. sapientum*	Mesocarp + endocarp
Cashewnut	*Anacardium accidentale*	Cotyledons and fleshy axis (Peduncle)
Coconut	*Cocus nucifera*	Endosperm and cotyledons
Cucumber	*Cucumis sativus*	Mesocarp, Endocarp, Placenta
Custard apple	*Annona squuamosa*	Fleshy Pericarp of individual Berry
Date fruit	*Phoenix sylvestris*	Pericarp (Mesocarp is fleshy)
Fig	*Ficus* sp.	Fleshy Receptacle
Grape	*Vitis vinefera*	Pericarp + Placenta
Guava	*Psidium guajava*	Pericarp + Thallamus
Jack fruit	*Artocarpous heterophyllus*	Bract, Perianth and seed
Lemon	*Citrus limon*	Juicy unicellular hairy out Growth from the inner wall of the endocarp.
Litchi	*Litchi chinensis*	Fleshy aril
Mango	*Mangifera indica*	Mesocarp
Melon		Mesocarp
Orange	*Citrus sinensis*	As in lemon
Papaya		Mesocarp
Peach	*Prunus persica*	Epicarp + Mesocarp
Pear	*Pyrus communis*	Fleshy thalamus
Pineapple	*Annas sativus*	Fused fleshy rachis, bract, perianth and pericarp.

Name of fruit	Scientific name	Edible parts
Plum	*Prunus domestica*	Epicarp + Mesocarp
Pome granate	*Punica granatum*	Succulent testa (juicy outer coat of seed)
Strawberry	*Fragaria* spp.	Succulent thalamus
Tomato	*Lycopersicon esuculentum*	Pericarp and placenta

Table 2: Edible parts of different vegetables used for culinary purpose.

Edible parts	Vegetables
Whole plant	Radish, Winged bean
Root	Radish, Carrot, Turnip, Beet root, Sweet potato, Tapioca (Cassava)
Stem	Asparagus, Knol khol (Kohlrabi)
Leaf with tender stem	Spinach, Spinach beet, Kale, Cabbage, Lettuce, Amaranthus, Fenugreek, Coriander, Celery, Parsley, Green onion and Garlic, Leek, Mustard, Swiss chard.
Flower and flower buds	Cauliflower, Broccoli, Artichoke
Fruit	Tomato, Brinjal (Eggplant), Chilli, Capsicum, Okra, Peas, Beans, Cucurbits
Seed	Peas, BeansH
Bulb (Dried) and Green bulb with leaves	Onion, Garlic, Shallot
Tuber	Potato
Rhizome	Ginger, Turmeric
Corm	Colocasia, Elephant foot

Various Uses of Vegetables and Fruits

Mostly ripe fruits are consumed as raw and used for processing for future consumption. Some unripe fruits are also used for processing such as preparation of pickles and dried powder. Whereas different Vegetables are utilized in various ways, almost all vegetables are used after cooking. But some of them are consumed as raw as well as in the form of salad (Table 3).

Table 3: Various uses of vegetables and fruits

Uses	Vegetables
Cooking	Most of vegetables are cooked and consumed with meal. Unripe jackfruit fruit is also cooked as vegetable.
Salad	These vegetables such as radish, carrot, turnip, tomato, lettuce, Onion, Chilli, cabbage, cucumber, Chinese cabbage, sweet Pepper are eaten as salad.

(*Contd.*)

Uses	Vegetables
Raw	Muskmelon, watermelon, cucumber, radish, carrot, turnip, tomato, All fruits when ripe are eaten raw.
Garnishing	Coriander, parsley, mint.
Chantney	Tomato, onion, garlic, chilli, coriander, mint.unripe mango, Karonda, Coconut
Pickles	Radish, turnip, carrot, cauliflower, chilli, garlic, Green, Mango ftuits Karonda, Jack fruit, Tind, Aonla
Spices	Fenugreek, onion, garlic, chilli, shallot, coriander,
Medicines	Fenugreek, onion, garlic
Products development	
Fruit juice and pulp	Mango, Guava, Grape
Fruit Squash	Mango, Citrus, Phalsa
Jam	Mango, Apple, Sapota
Jelly	Guava
Preserves	Mango, Apple, Aonla,
Candies	Pumpkin, Carrot,Ber, Kauronda, White squash
Fruit leather (Amawat)	Mango
Amchur (dried fruit powder)	Mango, Ber
Chips	Unripe Banana fruits, Potato
Wine	Grapes

Horticulture is a major part and play important role in Agricultural field in the modern age. It has become as a business enterprise, as the economic development enhanced. Due to increasing awareness of economical and nutritional importance of fruits and vegetables, Horticulture industry has emerged as an important sector of agriculture and has established credit in the improvement of farm income through increased productivity, generation of employment and in enhanced exports besides providing nutritional security.

The seasonal surpluses of fruits and vegetables during certain parts of the year in different regions can be processed and preserved and they can be used when there is scarcity of fresh fruits and vegetables, which will reduce the wasted of large quantities surplus fruits and vegetables due to improper handling and absence of facilities storage, distribution and marketing.

Fruits and vegetables are highly perishables in nature due to which post-harvest losses start immediately after harvesting that cause quantitative and qualititative deterioration and fruits and vegetables produces become inconsumable. On the other hand there is scarcity of fruits and vegetables in off season. To prevent wastage and make more availability of fruits and vegetables to consumers care must be taken and the fruits and vegetables can be processed using simple technologies. There are many processing methods that can be used on small-scale such as drying, fermenting, canning, freezing, preserving and juicing. Dried fruits and vegetables can be stored for long tie and used or sold in the future. Processed products of fruits and vegetables must be packaged and stored properly in order to enhance shelf life of products. Dried products must be packaged in air-tight containers such as glass or plastic bottles or sealed plastic bags. Canned and bottled products must be adequately heat processed using high quality containers that provide good seals. Dried and canned or bottled products are best stored in a cool, dark place.

Postharvest handling, transport and marketing of processed products can be much simpler and less costly than for fresh products, since refrigeration is unnecessary. Dried products take up much less space than their fresh and reduce transport and storage costs.

It is reported that India is second largest producer of vegetables and fruits after China but consumption rate is low particularly in rural areas as compared to urban areas. It is estimated that per capita fruits availability in our country is 230 gms. per capita per day (Horticultural Statistics at a Glance 2017). However, there is increasing demand of fruits and vegetables at consumer level because of awareness of their nutritional roles in human health and cheaper and better source of protective food. If they can be supplied fresh or processed forms throughout the year for human consumption, the national picture regarding health will be improved greatly.

Therefore it is urgently needed to establish enough processing units and encourage people towards this field of horticulture.

2

Present Status of Fruits and Vegetables Production

World Scenario

The world climatic conditions are divided into Tropical; Sub-Tropical and Temperate climates. Fruits and Vegetables suited to these climatic conditions are grown successfully. Considering the nutritional and economic importance of fruits and vegetables, there is tremendous increase in area and production of fruits and vegetables throughout the world. The present global scenario of fruit and vegetable crops (areas harvested, production and productivity (yield) has been given in table 1. It shows that the fruits 865.88 million tones has been harvested from 65.24 million Ha land with productivity 13.27 t/ha (Table 1) where as vegetables produced 1075.20 million tonnes from area 57.00 million ha with yield 18.86 tonnes/ha (Table 1) during 2016.

Table 1: Major Fruits and Vegetables producing countries in the World 2016

Fruit (excl. Melon)				Vegetables & Melons			
Country	**Prod. (Million Tonnes)**	**Area (Million Ha)**	**Yield (T/Ha)**	**Country**	**Prod. (Million Tonnes)**	**Area (Million Ha)**	**Yield (T/Ha)**
China	272.08	16.59	16.40	China	543.16	23.55	23.07
India	90.89	6.98	13.03	India	119.91	8.20	14.63
Brazil	39.69	2.33	17.02	USA	33.12	0.99	33.62
USA	27.11	1.19	22.73	Turkey 2	24.40	0.73	33.41
Mexico	21.43	1.42	15.13	Iran	14.29	0.46	31.31
Spain	19.05	1.63	11.66	Egypt	16.80	0.68	24.62
Indonesia	18.52	0.77	24.17	Russian Fed.	16.29	0.68	24.03
Philippines	16.32	1.68	9.72	Mexico	14.33	0.71	20.07
Italy	18.00	1.14	15.79	Spain	13.60	0.31	44.28
Turkey	21.74	1.38	15.79	Italy	12.32	0.47	26.11
Others	320.62	28.60	11.21	Others	249.65	16.51	15.12
World total	865.88	65.24	13.27	World	1075.20	57.00	18.86

Source: Horticultural Statistics at a Glance 2018

According to report China stands on first in both fruits and vegetable production with 272.08 million tonnes and 543.16 million tonnes respectively followed by India with 90.89 million tonnes fruits and 119.91 million tonnes vegetables during the year 2016. Mexico had lowest production of fruits (21.43 million tonnes and Italy had lowest vegetable production only 12.32 million tonnes during the year 2016.

Country Scenario

India is having all types of agro-climatic conditions suitable for growing different tropical, sub-tropical and temperate fruits, vegetables and other horticultural crops. The development achieved in the horticulture sector indicates that there is a growing demand for horticulture produce and the scenario of horticulture crops in India has become very encouraging. There are several types of fruits and vegetables are available in market and to the consumers. It has also been realized that the development of horticulture has positive impacts on the environment and ultimately on human health.

Production

The horticulture production has tremendously within few years in India. According to the report of Horticultural Statistics at a Glance 2018 the total production of fruits and vegetables during the year 2014-15 was 89514 and 166566 thousand million tonnes respectively which increased to 97358 and 184394 thousands million tonnes during the year 2017-18 respectively (Table 2). Now India is the second largest producer of fruits and vegetables next to China in the World. Not only this India is a leader in production of many fruits and vegetables such as Okra, Banana, Mango, Mangosteen, Guava, Lemon and Lime, and Papaya. India also holds second ranks in production of several vegetables like Potato, Tomato, Onion (dry), Cabbage and Brassicas, and Brinjal or Eggplant in the world. The percent share of vegetable production is 11.2 and the percent share of fruit production is 10.5 (Table 3).

Table 2: Area and Production of Horticulture Crops for four years

Area in '000 Ha ; Production in '000 MT

Crops	2014-15		2015-16		2016-17		2017-18	
	Area	Production	Area	Production	Area	Production	Area	Production
Total Fruits	6235	89514	6301	90183	6373	92918	6506	97358
Total Vegetables	9417	166566	10106	169064	10238	178172	10259	184394
Aromatic Plants	659	1000	634	1022	664	972	720	866
Total Flowers	249	2143	278	2184	306	2392	324	2785

(Contd.)

Crops	2014-15		2015-16		2016-17		2017-18	
	Area	Production	Area	Production	Area	Production	Area	Production
Total Plantation	3534	15575	3680	16658	3598	17972	3744	18082
Total Spices	3317	6108	3474	6988	3671	8122	3878	8124
Total	23410	280986	24472	286188	24851	300643	25431	311714

Table 3: India's Position in World Horticultural Crops 2016

Item	India	World	India's		Next to
			% Share	Rank	
Fruits & Vegetables (Million Tonnes)					
Vegetables & Melons	120	1075	11.2	Second	China
Okra	5.5	9	62	First	
Potatoes	44	377	11.6	Second	China
Tomato	18.4	177	10.4	Second	China
Onion (dry)	19.4	93.2	21	Second	China
Cabbages & other Brassicas	9	71.2	12.3	Second	China
Cauliflower & Broccoli	8.2	25.2	32.5	Second	China
Brinjal	12.6	51.3	24.5	Second	China
Fruits excluding Melons	91	866	10.5	Second	China
Banana	29.1	113.2	25.7	First	
Mango, Mangosteen and Guava	18.8	46.5	40.4	First	
Lemon & Lime	3	17.3	17.2	First	
Papaya	5.6	12.6	44.4	First	

Consumption of Fruits and Vegetables

It is well known that fruits and vegetables are highly nutritious and play vital roles in human health and have important place in our daily diet. On the basis survey made by experts of FAO and WHO, the combine panel of both on diet, nutrition and prevention of chronic diseases has recommended a daily intake of at least **400 grams** of fruits and vegetables, excluding potatoes, cassava and other starchy tubers, to prevent diet related chronic diseases and micro-nutrient deficiencies (FAO/WHO Expert Consultation. 2003) (http://faostat.fao.org/site/609/DesktopDefault.apspx).

Similarly, National Nutrition Guidelines recommends average daily consumption of 300 g vegetables and 100 g fruits. The vegetables include green leafy vegetables = 50 gm, other vegetables = 200 gm, roots and tubers = 50 gm.(National Institute of Nutrition 2011). But according to FAOSTAT (2010) report the consumption of fruits and vegetables is 311.03 g/capita/day i.e., 123.44 g/capita/day fruits and 177.59 g/capita/day vegetables during 2007 which is very low as compared to FAOWHO (2003) recommendation (table 4).

Table 4: Consumption of fruits and vegetables in India during 2007*

Items	Quantity (g/capita/day)
Fruits excluding wine + (Total)	123.44
Vegetables + (Total)	177.59
Vegetables other	131.78
Vegetable products	No data available
Roots other	No data available

****Source:*** FAOSTAT (2010),
(http://faostat.fao.org/site/609/DesktopDefault.aspx?PageID609#ancor)

The consumption of fruits and vegetables in American, European and some Asian countries is more or not less than recommended amount. But there is low consumption of fruits and vegetables in India and mostly African countries. However, when we compare the consumption of fruits and Vegetables in India with some African countries like Ethiopia and Eritrea, the position of India is better than Ethiopia and Eritrea (Table 5). The consumption of fruits and vegetables in India is more (310.37) than these countries. In Ethiopia consumption of vegetables and fruits is about 97g/person/day, whereas the consumption of fruits and vegetables in Eritrea is only 28 g/capita/day (2.00 g of fruits and 26.00 g of vegetables). Though consumption of fruits and vegetables in India is more but it is very low as per recommendation given by FAO WHO (2003). The low consumption of fruits and vegetables in African countries is because in most African communities people rely on one or two staple crops such as maize, millet, sorghum, rice, teff, cassava yam, sweet potato, plantain and enset. The crops tend to provide the bulk (about 60 to 80 percent) of the energy intake of household members. They typically include roots and tubers, green leafy vegetables, legumes, and fruits. Fruits and Vegetables are important components of a healthy diet.

Table 5: Comparison of fruits and vegetables consumption in India with some African counties.

Vegetables/fruits	Consumption (g/capita/day)			
	FAO/WHO	India	Ethiopia	Eritrea
Fruits/Vegetables	400*	310.37**	97***	28****

Source: * http://faostat.fao.org/site/609/DesktopDefault.apspx
**FAOSTAT (2010) (http://faostat.fao.org/site/609/DesktopDefault.aspx?PageID609#ancor).
*** Other source
****http://faostat.fao.org/site/609/DesktopDefault.apspx

Projected Availability and Demand

According to a report projected per capita availability of fruit and vegetables in India is 207.9 g/capita/day and 393.76 g/capita/day respectively (Table 6). This is due to increased area and production of fruits and vegetables which might have been influenced by increased consumption due to awareness of nutritional importance of fruits and vegetables in the society. According to economic if there is more consumption ultimately demand will increase which will influence the supply and production will also increase. Therefore, different fruits and vegetables will be available in adequate quantity.

Table 6: Per capita availability of Fruits and Vegetables

Projected Population of India in 2017-18 (million)	1283
Fruits Production in 2017-18 (thousand tonnes)	97358.0
Vegetables Production in 2017-18 (thousand tonnes)	184394.0
"Per Capita Availability of Fruits (gms/person/day)"	207.9
"Per Capita Availability of Vegetables (gms/person/day)"	393.76

Source: Population Projections for India and States, 2001-2026 Report, Office of Registrar General of India (RGI), May 2006 (Horticulture at Glance 2018).

India is the second most populated country in the world with over 1.21 billion people (2011,census). It is projected to be the world's most populated country by 2025, surpassing China and its population will reach 1.6 billion by 2050 (BBC News. 18-08-2004). With increasing population, the demand for horticulture produce will also increase. Thus it has been projected that the country will require 360 million tonnes of horticulture produce by 2020-21 (Table 7).

Table 7: Projected Demand of horticultural crops*

Commodity	**Production (Million tones)**		
	1991-92	**2011-12**	**2020-21**
Fruits	28.63	81.00	98.00
Vegetables	58.53	185.00	220.00
Spices	1.90	5.50	650.00
Coconut	7.49	20.00	24.00
Cashewnut	0.40	1.70	2.00
Cocoa other	-	6.80	9.50
Total	96.95	300.00	360.0

*****Source:*** Singh, H. P. (2007), Indian Horticulture, 52(4); N H B (2009) cited by Singh and Malhotra (2011).

Export and Import of Fresh and Processed Fruits and Vegetables

There is tremendous increase in production of fruits and vegetables and India has second rank after China which has offers great opportunities for export and during 2018-19 the country has exported fruits and vegetables worth ₹10,236.93 crores/1,469.33 USD Millions which comprised of fruits worth ₹4,817.35 crores/692.01 USD Millions and vegetables worth ₹5,419.48 crores/777.25 USD Millions. Major contribution of fruits such as Grapes, Pomegranates, Mangoes, Bananas, Oranges and vegetables Onions, Mixed Vegetables, Potatoes, Tomatoes, and Green Chilli is in Indian export. The major countries importing fruits and vegetables from India are Bangladesh, UAE, Netherland, Nepal, Malaysia, UK, Sri Lanka, Oman and Qatar (APEDA, http://apeda.gov.in/apedawebsite/six_head_product/FFV.htm).

Besides the export, country is also importing fresh and processed vegetables every year due to increasing demand to fulfill the requirement of people. During 2017-18 country has imported fresh fruits and vegetables worth ₹1,255,016.61 lacs and processed fruits and vegetables worth ₹93,718.56 lacs (Table 8).

Table 8: Import of fresh and processed fruits and vegetables for three years in India

Qty in MT; Value in Rs. Lacs

	2015-16		2016-17		2017-18	
	Qty	**Value**	**Qty**	**Value**	**Qty**	**Value**
Fresh	980271.00	1146601.60	1048743.00	1125207.10	1010357.00	1255016.61
Processed	54341.53	64682.10	56112.67	66418.21	68636.93	93718.56

Source: APEDA Horticultural Statistics at a Glance 2018

Scope of Fruit and Vegetable Production

There is great scope to exploit full potentials of horticultural crops particularly fruits and vegetables to fulfill the demand of the country as well as Global market. There are several conditions which increase the scope of production of fruits and vegetables.

(i) Climatic conditions: There are suitable climatic conditions such as temperate, sub tropical and tropical climates available in different parts of India. Suited fruits and vegetables can be grown in different climatic conditions successfully.

(ii) Increasing demand: Nutritionally fruits and vegetables are very important and they provide all nutrients such as carbohydrates, proteins, fats, minerals and vitamins either in small or large quantity required by our body to maintain proper health. Not only this they are also richest and cheapest source of protective food. Besides nutrients they also contain phytochemical as antioxidants which provide power to fight against adverse environmental conditions. The fruits and vegetables are rich in fibres which play vital role in our digestion. Due to the awareness of nutritional importance of these food commodities there is increasing the demand at consumer level. This will certainly influence the scope increases the areas and production of fruits and vegetables to fulfill the requirement of the consumers.

(iii) High market price: According to economic the increase demand and low supply of commodity always increases market price. This will certainly compel farming community to grow more fruits and vegetables because fruits and vegetables are high yielders as compared to other crops and also have high market price. Growing of fruits and vegetables has become source of continuous income to the farming as well trading communities.

(iv) Food security: Fruits and vegetables being important food commodity play very important role in food security. Some fruits and vegetables have longer shelf life and storage life and can be stored for longer period. These fruits and vegetables can be used as food during the scarcity period of food grains. Banana, Enset, Potato, Sweet potato and Tapioca play very important role in food security. Banana and Enset are most important fruits used during scarcity of food when there is draught period and grains crops production is nothing. Among vegetables potato, sweet potato and tapioca or cassava have longer storage life and can be stored for long period. They can be used during the time when food grains are not available.

(v) Poverty alleviation: When fruits plants are once established, they continue to bear fruits for several years continuously. Vegetables are high yielder and short duration crops and can be grown three times in a year. Fruits and vegetables being high yielders can be harvested several times. That can generate continuous income to the farmers. High productivity and high market price help both farmers and poor traders to achieve more benefit to alleviate the poverty.

(vi) Alleviation of malnutrion: There is serious problem of malnutrion in many counties. Mostly children are suffering this malnutrition. Malnutrion may cause disease like Rickets and Night blindness. It is also responsible for poor health. There are several cases of death of children due to malnutrition. Malnutrion can be alleviated by right and recommended amount of fruits and vegetables in diet.

To alleviate the poverty and malnutrition in the rural people, the only solution is to promote horticultural crops such as fruit and vegetable production.

(vii) Export potential: The export of fresh fruits and vegetables is currently becomes very wide due to WTO and there is great demand of fruits and vegetables in the international markets which increases the good prospects towards the diversification of exports and earning foreign exchange. There are several fruits and vegetables having export potential but the export potential of these fruits and vegetables are not exploited. When export potential of fruits and vegetables will increase ultimately cultivation and production of fruits and vegetables will increase.

3

Importance of Fruits and Vegetables in Human Nutrition (A Nutritional Approach)

The food we eat contains nourishing substances called nutrients which are required by human body for proper functioning and maintenance of body health. They are classified into five groups such as:

(i) Carbohydrates,
(ii) Protein,
(iii) Fats,
(iv) Minerals
(v) Vitamins

All these nutrients required by our body are supplied by following foods commodities.

(i) Grains
(ii) Vegetables
(iii) Fruits
(iv) Meat, fish and beans (Meat, poultry, fish, dry beans, eggs, nuts, and meat are alternatives)
(v) Milk (which includes yogurt and cheese).

The food we eat contains nourishing substances called nutrients required by human body for proper functioning. They are classified into five groups such as (i) carbohydrates, (ii) protein, (iii) fats, (iv) minerals and (v) vitamins. First three nutrients are required in large quantity and all of them supply energy to the body. There is high requirement of energy by adult man and woman body which is supplied by carbohydrates, protein and fats. Vitamins and Minerals are required in small quantity and essential for physiological processes and metabolic activities. These essential nutrients can be grouped into three groups according to their functions in the body.

A. **Body builders** : Proteins
B. **Energy Producing** : Carbohydrates, Fats and up to some extend from proteins
C. **Protective food** : Vitamins and Minerals

Latham (1997) has described the chemical composition of human body (weight 65 kg) in his book "FAO Human Nutrition in the Developing World" FAO. Rome".

Components	Percentage of body weight
Water	61.6
Proteins	17
Fats	13.8
Minerals	6.7
Carbohydrates	1.5

He has also mentioned simple classification of dietary constituents given below.

Constituents	Uses
Water	To provide fluid and to help regulate body temperature
Carbohydrates	As fuel for energy for body heat and work
Fats	As fuel for energy and essential fatty acids
Proteins	For growth and repair
Minerals	For developing body tissues and for metabolic processes and protection
Vitamins	For metabolic processes and protection
Indigestible and unabsorvable	For a vehicle for other nutrients, add bulk to diet, provide a particles including fibers habitat for bacterial flora and assist proper elimination of refuse.

Source: Latham M.C. (1997) "FAO Human Nutrition in Developing World", FAO, Rome, 1997.

Nutrients requirement of human body

The Carbohydrates, Protein and Fat are needed in large quantity but Minerals and Vitamins are required in small amount but they are very essential for human body. The recommendations of different nutrients required per capita per day by our body has been described by Gupta and Gupta (1986) presented in the Table 1.

Table 1: Daily requirement of nutrient*

Nutrients	Quantity required	Nutrients	Quantity required
Carbohydrate	100g	Vitamin "C"	50 mg
Proteins	45-60 g	Vitamin "D"	2.5 micro gram
Fats (vegetable oil)	15g	Vitamin "E"	30 mg
Fibres	4-5 g	Vitamin "K"	140 micro gram
Energy	1900-3900 Calories	Minerals	
Vitamins		Calcium	500 mg
Vitamin "A"	750 micro gram	Phosphorus	800 mg

(Contd.)

Nutrients	Quantity required	Nutrients	Quantity required
Vitamin "B"	Complex	Iron	20 mg
Thiamine (B1)	1.2 mg	Iodine	0.2 mg
Niacin	6.6 mg	Magnesium	400 mg
Riboflavin	1.7 mg	Sodium	1100-2000 mg
Pyridoxine (B6)	1.25 mg per 100g of protein consumed	Zinc	5-10 mg
Pantothenic acid	10 mg Copper	3 mg	
Biotin	1 mg	Potassium	1400-2000 mg
Cyanocobalamin (B12)	2 mg	Water 1.5-2.0 litres	
Folic acid	200 micro gram		
Choline	10 mg		
Inositol	10 mg		

****Source:*** Kusum Gupta and L.C. Gupta (1986) "Food and Nutrition" Published by Jaypee Brother's Medical Publishers, New Delhi, on page 52

Dietary allowances for adult man and adult female

Adult man and woman require more energy and the need of calories and nutrients per person per day for adult man and woman recommended by Indian Council of Medical Research (ICMR) presented in Table 2.

Table 2: Minimum amount of Calories and nutrients required per caput per day for adult male and adult Woman

Calories/Nutrients	Amount required per caput per day	
	Adult Male	Adult Femal
Calories(Kilocalories)	2800	2200
Proteins (g)	55	45
Calcium (mg)	400-500	400-500
Iron (mg)	20	30
Carotene (microgram)	3000	3000
Thiamin (mg)	1.4	1.1
Riboflavin (mg)	1.5	1.2
Ascorbic acid (mg)	50	50
Folic acid (microgram)	100	100
Vitamin B12 (microgram)	1.0	1.0

****Source:*** ICMR (Indian council of Medical Research) Cited by Vishnu Swarup (2006) "Vegetable Science and Technology in India", Kalyani Publishers, Ludhiana, New Delhi, Noida,(U. P.), Hyderabad, Chennai, Kolkata, Cuttack, Guwahati. First Edition, 2006.

Fruits and vegetables are considered as nutritionally essential food commodity and due to high nutritive value, fruits and vegetables are important ingredients in human diet. The roles of fruits and vegetables in human health have been described below.

(1) Fruits and Vegetables rich sources of nutrients

Fruits and vegetables contain almost all nutrients required by our body and supply either in small or large quantity. As a food group, fruits and vegetables significantly enhance the quality of the human diet by contributing variability, bulk, fibers, vitamins, minerals and other compounds, which have positive effect on human health.. Besides these nutrients some fruits and vegetables contain dietary fibres which play very important role in our diet but it is not considered as a nutrient because it is not absorbed by blood stream.

Fruits and vegetables are richest and cheapest source of all nutrients (carbohydrates, proteins fat, minerals and vitamins) which are required for our body to maintain good health and proper functioning of body organs. Besides they are also rich in water and dietary fibers required for the human body. Some vegetables can be grouped according to the primary type of nutrient they provide given below (Rubatzky and Yamaguchi, 1999).

Carbohydrate: White potato, Sweet potato, Dry bean, Cassava, Yam, Taro, Plantain.

Fat: Mature seed of some Legumes and Cucurbits.

Protein: Beans, Peas, Sweet corn, Leafy crucifers.

Pro-Vitamin A: Carrot, Orange/yellow flesh Sweet potato, Squash, Green leafy vegetables, Pepper, Green beans, Green peas.

Vitamin C: Crucifers, Immature bean seeds, Tomato, Bean sprouts, Pepper, White potato, Melons, Many leafy vegetables.

Minerals: Crucifers, Many other leafy vegetables.

Source: Rubatzky, V.E. and Mas Yamaguchi, (1997) World Vegetables Principles, Production, and Nutrition, A Chapman & Hall Food Science Book, An Aspen Publication.

Different fruits and vegetables rich in nutrient content are given below (Table 3):

Table 3: Nutrients rich fruits and vegetables

Nutrients	Vegetables/Fruits
A. Vegetables	
Carbohydrate	White potato, Sweet potato, Dry bean, Cassava, Yam, Taro, Plantain.
Fat	Mature seed of some Legumes and Cucurbits.
Protein	Beans, Peas, Sweet corn, Leafy crucifers.
Pro-Vitamin A	Carrot, Orange/yellow flesh Sweet potato, Squash, Green leafy vegetables, Pepper, Green beans, Green peas.

(Contd.)

Nutrients	Vegetables/Fruits
Vitamin C	Crucifers, Immature bean seeds, Tomato, Bean sprouts, Pepper, White potato, Melons, Many leafy vegetables.
Vitamin B1 (Thiam)	Dry peas and Beans
B2 (Riboflavin)	Green leafy vegetables
Vitamin B complex (B6 and B12)	Peas, Beans, Garlic, Colocasia, Tomato, Lettuce, Asparagus, Cabbage,.
Calcium:	Kale, Turnip greens, Broccoli, Dried Bean and Peas, Spinach, Carrot, Onion, Beet root, Coriander, Tomato, Cauliflower, Pumpkin, aranthus, Fenugreek, Drumstick leaves, Lettuce, Cabbage, Swiss chard, Mustard leaves
Phosphorus:	Beans, Potato, Carrot, Tomato, Cucumber, Lettuce, Garlic, Peas, Cauliflower, Cabbage, Beet root
Potassium	Potato, Tomato, Sweet potato, Bitter gourd, Radish, Legumes, Winter squash
B. Fruits	
Carbohydrates	Date, Banana, Apricot (dry)
Protein	Avocado, Papaya and Nuts (Almond Cashew nut Walnut)
Fat	Nuts and Avocado
Vitamins	Mango, Papaya, Fig
Vitamin B1 (Thiamin)	Banana, Apricot,Aapple, Pear, Plum and Nuts
Vitamin B2 (Riboflavin)	Papaya, Pear, Pomegranate
Vitamin B complex (B6 and B12)	Peach, Apple, Guava, Grapes, Pineapple, Pomegranate and Nuts
Calcium and phosphorus	Nuts and Banana
Potassium	Banana, Blue berry

Nutritional values of some major fruits are given below (Table 4) which show that Avocado is richest source of nutrients and Avocado, Plantain, Banana are richest in energy and protein. Whereas, orange is richest in calcium. Plantain is richest in Iron. Mango and Plantain are richest in Vitamin A and Guava is richest source of Vitamin C.

Leafy vegetables and root and tubers crops are nutritionally more important to be consumed. Because root and tuber vegetables are rich in carbohydrates and supply large amount of energy. Nutrients contents of some leafy vegetables and root and tuber crops are given below in Table 5 and 6.

Table 4: Nutrients of fruits in 100g edible portion*

Food (Waste %) (Fruits)	Energy (Kcal)	Protein (g)	Fat (g)	Calcium (mg)	Iron (mg)	Vit A (g)	Thiamin (mg)	Riboflavin (mg)	Niaci-n (mg)	Folate (Folic acid) (mg)	Vit C (mg)
Avocado (50)	161	20	15.3	11	1.02	61	0.11	0.12	1.9	22	8
Banana (33)	92	10	0.5	6	0.30	20	0.05	0.16	0.5	19	9
Orange (28)	47	0.9	0.1	40	0.10	120	0.09	0.09	0.3	30	53
Lime (36)	30	0.7	0.2	33	0.60	1	0.03	0.02	0.2	8	23
Lemon (36)	29	0.6	0.3	26	0.60	3	0.04	0.02	0.1	11	53
Guava (11)	51	0.8	0.6	20	0.30	79	0.05	0.06	1.2	7	184
Mango (31)	65	0.5	0.3	10	0.10	389	0.06	0.06	0.6	7	28
Papaya (46)	39	0.6	0.1	24	0.10	201	0.03	0.03	0.3	1	62
Water melon (56)	32	0.6	0.4	8	0.20	37	0.08	0.02	0.2	2	10
Plantain (34)	135	1.2	0.3	8	1.30	390	0.08	0.04	0.6	16	20
Pineapple (46)	49	0.4	0.4	7	0.40	2	0.09	0.04	0.04	11	62

Source: *Latham M.C. (1997) in his book “FAO Human Nutrition in Developing World”, FAO, Rome, 1997.

Table 5: Nutritive values of Leafy vegetables (Per 100 g of edible portion)*

Leafy vegetable	Moisture G	Energy Kcl	Protein G	Fat G	Ca Mg	Fe Mg	Carotene Mcg	B1 Mg	B 2 Mg	Niacin Mg	Vit. C mg
Agathi	73.1	93	8.4	1.4	1130	3.9	5400	0.21	0.09	1.2	169
Amaranth	85.7	45	4	0.5	397	3.5	5520	0.03	0.30	1.2	99
Cabbage	91.9	27	1.8	0.1	39	0.8	120	0,06	0.09	0.4	124
Curry leaves	63.8	108	6.1	1.0	830	0.9	7560	0.08	0.21	2.3	4
Coriander leaves	86.3	44	3.3	0.6	184	0	6918	0.05	0.06	0.8	135
Drumstick leaves	75.9	92	6.7	1.7	440	0.8	6780	0.06	0.05	0.8	220
Fenugreek leaves	86.1	49	4.4	0.9	395	1.9	2340	0.04	0.31	0.8	52
Spinach	92.1	26	2.0	0.7	73	1.1	5580	0.03	0.26	0.5	28

Table 6: Nutritive values of Root and Tuber vegetables (Per 100 g of edible portion)*

Vegetables	Energy Kcl	Protein g	Carbohydrate g	Ca mg	Fe mg	Carotene mcg	B 1 mg	B 2 mg	Niacin mg	Vit. C mg
Beet root	43	1.7	8.8	18.3	1.19	O	0,04	0.09	0.4	10
Carrot	48	0.9	10.6	80	1.03	1890	0.04	0.02	0.6	3
Colocasia	97	3	21.1	40	0.42	24	0.09	0.03	0.4	0
Onion big	50	1.2	11.1	46.9	0.60	0	0.08	0.01	0.4	11
Onion small	59	1.8	12.6	40	1.2	15	0.08	0.02	0.5	2
Potato	97	1.6	22.6	10	0.48	24	0.1	0.01	1.2	17
Radish Pink	32	0.6	6.8	50	0.37	3	0.06	0.02	0.4	17
Sweet potato	120	1.2	28.2	46	0.21	6	0.08	0.04	0.7	24
Tapioca	157	0.7	38.1	50	0.9	--	0.05	0.10	0.3	25
Yam	111	1.4	2.6	35	1.19	78	0.07	--	0.7	--

* ***Source:*** Nutritive value of Indian Foods, National Institute of Nutrition (NIN), Hyderabad, India Cited by Nandini, P. V. (2005) ICAR Winter School on "Trade oriented exploitation of Horticulture in Humid Tropics- Opportunities and Challenges", 1-21 December 2005.

(2) Fruits and vegetables are rich in Phytochemicals (Antioxidants)

Certain vitamins and minerals such as vitamin E, vitamin C, zinc and selenium are believed to have antioxidant effects. Other plant-derived molecules include polyphenolic compounds also participate in these activities. This family is made up of primarily the following substances (L. Depezay–Fondation Louis Bonduelle–September 2007).

(i) **Anthocyanins,** natural colorants in red-violet vegetables (e.g. red onions, red cabbage),

(ii) **Flavones,** in cabbage and onions

(iii) **Phyto-oestrogens,** found in all vegetables

Wargovich (2000) reported that some components of fruits and vegetables are strong antioxidants and function to modify the metabolic activation and detoxification/disposition of carcinogens, or even influence processes that alter the course of the tumor cell.

The antioxidant capacity varies greatly among fruits and vegetables. It is better to consume a variety of commodities rather than limiting consumption to a few with the highest antioxidant capacity. (Importance of Fruits, Nuts, and Vegetables in Human Nutrition and Health Adel Kader, Department of Pomology, UC Davis May 2001, Issue No. 106 Perishables Handling Quarterly (http://ucanr.edu/datastoreFiles/234-104.pdf).

(3) Fruits and vegetables are rich in dietary fibres

Besides essential nutrients fruits and vegetables are also rich in dietary fibres which play important role digestion and help in the movement of food materials after digestion in alimentary canal. Thus they prevent constipation. Dietary fibres are also useful in prevention of certain chronic diseases like diabetes and blood circulatory diseases. Among fruits and vegetables rich in fibers, most vegetables particularly leafy vegetables, root and bulb vegetable crops and okra contain more amount of fibres. Although fiber play very important role in our diet but it is not considered as a nutrient because it is not absorbed by Blood Stream. These fibres are made up of Pectin, Gums, Cellulose, Hemicelluloses, and Lignin. The enzymes of stomach and small intestine do not digest dietary fires. But fibres have property of holding water and swell. Thus they add bulk to diet and increase transit time in the gut. Fibre containing vegetables yield a spongy mass on cooking which satisfy the appetite and assist in the smooth movement of food in the digestive tract.

(4) Fruits and vegetables provide water to the human body

Most fruits and vegetables particularly leafy vegetables such as Spinach, Cabbage, Kale, Swiss chard, Mustard leaves are succulent in nature and containing high amount of water about 80-95 percent. When these vegetables are consumed, they provide water to body. It is well known that water play very important role in the human body.

Role of water in human body

1. Water is required for maintaining osmotic balance in the cells
2. Water is needed to carry out metabolic processes and all chemical reactions in body take place in presence of water
3. Water is a constituent of all body fluids.
4. Water helps to maintain body temperature.
5. All products of digestion are dissolved in water and absorbed and transported throughout the body.
6. It is also required to carry out the metabolic waste products such as urine and swet from the body

(5) Place of fruits and vegetables in balanced diet

We eat five basic food groups and each food group provides certain nutritional benefits, therefore, foods from each group should be consumed each day. The key to a balanced diet is to recognize that grains (especially whole grains), vegetables and fruits are needed in greater proportion than foods from the meat, fish & beans and milk groups. A healthy and balanced diet also contains a variety of foods from each food group. Since each food offers different macronutrients (energy providing carbohydrates, proteins, and fats) and micronutrients (vitamins and minerals). Suitable combination of balanced diet is given below (Table 7).

Table 7: Suitable combination of balanced diet*

Food	Quantity (g)
Cereals	400
Milk and milk product	285
Green leafy vegetables	114
Root vegetables	85
Other vegetables	85
Fruits	85
Pulses, nuts and oil seeds	85
Fish and meal	85
Sugar	57
Vegetable oil Egg	57
Egg	40 (1 egg)

****Source:*** Fruit Growing by J.S. Bal, Kalyani Publishers, 2005.

The nutrition requirement of children is more important in which fruits and vegetables have important place. To satisfy the nutrient needs of children is shown in Table 8.

Table 8: Amount of uncooked foods to satisfy the nutrient needs of children (g)

Age in year	Cereal grain	Legumes	Vegetables	Fruits	Oil or Fat
2-3	15-250	100-125	75-100	50-100	20
4-5	200-350	125-175	100-150	100-150	30
6-9	300-400	150-200	100-150	100-150	30
10-13	400-500	200-250	100-150	100-150	30

****Source:*** FAO Human Nutrition in the Developing world by M.C. Latham, FAO, Rome 1997

Considering the role of fruits and vegetables in balanced diet, FAO WHO (2003 based on survey by their experts recommended 400 g/capita/day of fruits and vegetables for proper health. According to nutrition experts on individual should consume at least 100 g fruits and 300 g vegetables per day in addition to pulses, milk, meat, egg etc to maintain proper health..

(6) Fruits and vegetables act as protective food

Fruits and vegetables are rich in vitamins and minerals which provide strength to human body to resist against adverse conditions especially against disease organisms. Therefore, fruits and vegetables are called protective foods. Among vegetables leafy vegetables are richest source of minerals and vitamins (Table 5).

(7) Fruits and vegetables serve as subsidiary food

Fruits and Vegetables like Banana, Enset, Potato, Sweet potato, Tapioca and lesser Yam can be used as substitute for cereals like wheat and rice and they can also act as subsidiary foods.

(8) Fruits and vegetables increase appetite

Fruits and vegetables increase our appetite in many ways'

i. When eaten as salad with meal
ii. When cooked as vegetable and eaten with food.
iii. When cooked with food like pulaw (Rice pulaw).

In these forms they also add flavor and taste to our meal so that it is eaten more.

B. Medicinal Importance of Fruits and Vegetables

The fruits and vegetables play very important role in the balanced diet of human which supply energy rich food and vital protective nutrients like minerals and vitamins. Thus they are one of the cheapest sources of natural nutritive food and enrich human health. All these nutrients have positive impact on human health (Table 9). Low consumption of these protective food fruits and vegetables is largely responsible for malnutrition in human beings especially children.

Table 9: Positive impact of constituents of fruits and vegetables on human health and their sources (Produce for Better Health Foundation, 1999; USDA, 2000)*.

Constituent	Sources	Impacted human diseases
Antioxidants		Cancer, cataracts, heart disease, stroke
Vitamin C	Broccoli, cabbage, cantaloupe, citrus fruits, guava, kiwifruit, leafy greens, pepper, pineapple, potato, strawberry, tomato	
Vitamin A	Dark-green vegetables (such as collards, spinach,(carotenoids) and turnip greens), orange vegetables (such as carrots, pumpkin, and sweet potato), orange-flesh fruits (such as apricot, cantaloupe, mango, nectarine, orange, papaya, peach, persimmon and pineapple), tomato	
Vitamin E	Nuts (such nuts (such as almonds, cashew nuts, filberts, macadamias, pecans, pistachios, and walnuts)	
Flavonoids	Red, blue, and purple fruits (such as apple, blackberry, blueberry, cranberry, grape, nectarine, peach, plum & prune, pomegranate, raspberry, and strawberry)	
Fiber	Most fresh fruits and vegetables, nuts, cooked dry beans and peas	Diabetes, heart disease
Folate	Dark-green leafy vegetables, cancer, mustard greens, (such as spinach and romaine lettuce), legumes (cooked dry beans and peas, green peas), oranges	Birth defects, heart disease
Potassium	Baked potato or sweet potato, cooked dry beans, banana & plantain, cooked greens, dried fruits (such as apricots and prunes), winter (orange) squash	Hypertension, stroke

****Source:*** Kedar (2001), Importance of Fruits, Nuts, and Vegetables in Human Nutrition and Health by Adel Kader, Department of Pomology, UC Davis A pdf file May 2001, Issue No. 106 Perishables Handling Quarterly Page 5).

Those who consume little fruits and vegetables suffer from diseases caused by minerals and vitamins deficiency (Table 10).

Table 10: Consequences of Nutrient Deficiency*

S. No.	Nutrients	Consequences of deficiency
1.	Calories and Proteins	Retard growth in children, irritability, apathy and possibly retarded mental development, discolouration of skin and hair, welling of the face and lower part of the legs and feet, fatty liver, etc. (medically known as kwashiorkor) and extreme emaciation (marasmus).
2.	Vitamin A	Inability to see in twilight, sensivity to bright light, foamy white patches on the conjunctive (bitot spots), softening of the cornea, eventually leading to blindness and frequent respiratory infections.
3.	Vitamin B-Complex	
	(a) Thiamin (B1)	Causes beriberi disease and loss of appetite
	(b) Riboflavin (B2)	Cracks at the corners of the mouth, raw-red cracked lips, glossy tongue and ulcers in the oral cavity.
	(c) Nicotinic Acid (B5)	Sore tongue (scarlet coloured), "Pellagra" showing skin changes in hands, feet, legs, and neck.
	(d) Pyridoxin (B6)	Ulceration of oral cavity and associated with anaemia.
4.	Vitamin C	"Scurvy"-bleeding gums and mucus membranes and susceptibility to infection as common cold.
5.	Calcium	Important for bones and teeth, blood clotting, osteomalacia in women after repeated pregnancies.
6.	Iron	Anaemia-pale smooth tongue, pale lips, eyes and skin, spoon-shaped nails and frequent exhaustion.

****Source:*** Anonymous, 1986, In: Nutrition garden, Extension Bulletin 22, Indian Institute of Horticultural Research, Bangalore, India.

Fruits and vegetables are important components of a healthy diet, and their sufficient daily consumption could help prevent major diseases, such as cardiovascular diseases and certain cancers. A recently published WHO/FAO (2003) report recommends as a population-wide intake goal the consumption of a minimum of 400g of fruit and vegetables per day (excluding potatoes and other starchy tubers) for the prevention of chronic diseases such as heart disease, cancer, diabetes and obesity, as well as for the prevention and alleviation of several micronutrient deficiencies, especially in less developed countries. Research indicates that various phytochemicals found in fruits and vegetables are required for proper long-term cellular health and disease prevention. Regular consumption of fruit and vegetables is associated with reduced risk of cancer, cardio-vascular diseases (especially coronary heart disease), stroke, cataracts and some functional declines associated with aging (Liu, 78 (3): 5173-American Journal of Clinical). The indigestible fiber content of fruits and vegetables relieve or prevent constipation. Diets that are high in fiber are associated with lower body weight. Asthmatic individual who consume fruits and vegetables more than once a day having better lung function increase the ventilation function of the lungs. As fruits and vegetables contain several phytochemical compounds such as Carotenoides and Flavonoids that are antioxidants. They neutralize the adverse effects of free radicals produced in human body which are responsible early aging.

Most of vegetables have high medicinal values and recommended to be used as remedy of many diseases as described by Rubatzky and Yamaguchi (1997) shown in Table 11.

Table 11: Medicinal uses of vegetables

Vegetables	Uses for remedy of disease
Onion and Garlic	Headache, Parasites and Heart and Circulatory diseases, High blood pressure, Tumers, Cancer, Rheumatic disorders, Diabetes
Celery	Leprosy and some skin wounds.
Lettuce	Fevers, Headache and Jaundice, Urinary problems
Artichoke	Gastric disorders
Cabbage juice	Antidote for poisonous Mushrooms
Mustard	Reduce the pain of muscle strain or bruises.
Broccoli	Block the growth of tumers (Cancer)
Spinach extracts	Clean wounds and cure warts
Sweet potato	Lower fever, leaf tissues used to treat skin diseases
Bitter gourd (dried fruits)	Rheumatism and parasites, used as antitumor compound and skin disorder and burn.
Watermelon (seed)	Urinary tract infection, poor blood circulation
Cucumis melo (see Asparagus (shoots)	Used to cure Jaundice
Rhubarb	Liver diseases and stomach disorder
Pepper	Reduce symptoms of psoriasis, Arthritis and Contact allergy

Source: Rubatzky, V. E. and Mas Yamaguchi, (1997) World Vegetables Principles, Production, and Nutrition, A Chapman & Hall Food Science Book, An Aspen Publication

All of the hungry and many of the overweight are afflected with micronutrient deficiency (lack of vitamins and minerals). Horticultural crops can play a vital role in solving this global micronutrient crisis. Vegetables and fruits are the most sustainable and affordable sources of micronutrients in diets (UN, 2004). Healthy diets improve the learning capacity of children and the productivity of workers Vitamin A deficiency alone weakens the immune system of 40% of children in developing countries (UN, 2004). This deficiency increases a child's risk of severe illness and death from infectious diseases, which are the leading causes of death in developing countries. Deficiencies of vitamin A and other micronutrients may increase the likelihood of HIV/AIDS transmission from mother to child and hasten the progression of the disease in infected persons. Iron deficiency affects at least 2 billion, and perhaps up to 3.5 billion people. Shortage of iron in the diet causes reduced productivity, which results in economic losses of billions of dollars globally (UN, 2004).

C. Improved Food Security and Nutrition

Being rich in minerals and vitamin, fruit and vegetable crops assume great importance in nutritional security of masses. Fruits and Vegetables such as Banana, Enset, Sweet potato, Tapioca and Potato etc play vital roles in the Food and Nutritional security to the Nation.

In some countries these are used as staple foods. Even if fruits and vegetables are consumed daily with meal, they compensate the quantities of other food items, which can be saved for future. Fruit crops are perennial and do not require much cares and irrigation. They can survive even under dry i.e. arid and semi-arid conditions. Fruit like Banana and Enset play important role in National food security. In some countries they are used by rural people during difficult time when there is shortage of staple foods. Keeping above views in mind I can say:

"Eat More Fruits and Vegetables and Keep Body Healthy"

Fruits and vegetables arranged in shop

4

Post-harvest Losses and Controls in Fruits and Vegetables

Fruits and Vegetables are highly perishable in nature due to which huge spoilage (damage) takes place during the period between harvesting and consumption. Fruits and Vegetables after harvesting are often poorly handled and in many cases neither cleaned nor washed. Improper handling may cause physical damage due to tissue breakdown which results mechanical damages and physiological changes in harvested fruits and vegetables. Mechanical losses include bruising, cracking, cuts, microbial spoilage by fungi and bacteria and physiological losses include changes in respiration, transpiration, pigments, organic acids and flavour.

Extent of Post-Harvest Losses

There are many sites where post-harvest losses are noticed:

i. Farmer's field
ii. Packaging
iii. Transportation
iv. Marketing

Post harvest losses in fruits and vegetables vary 20-40% and about 10-15% fresh fruits and vegetables shrivel and decay which causes lowering their market value and consumer acceptability. In the present situation where per capita consumption of fruits and vegetables is still very low, reducing the post-harvest losses of fruits and vegetables is of vital importance. Minimizing these losses can increase their supply and availability to the consumers. In harvesting season, high production of fruits and vegetables results in a glut in the market and become scarce during other seasons. Surplus production of fruits and vegetables neither can all be consumed in fresh condition nor sold at economical prices. But unfortunately both fruits and vegetables are highly perishable in nature and there is great loss in their physico-chemical properties after harvesting due to absence of infrastructure and storage facilities that results wastage of fruits and vegetables (Table 1). Food grains have been taken care of only. Whereas fruits and vegetables have suffered post harvest losses for want of a simple technology of processing, preservation and transport and only about 1 % of the total fruits and vegetables produced are processed. According to recent report of CIPHET 2015 given in Horticultural Statistics at a Glance (2018), the maximum post-harvest losses 15.88% found in Guava and minimum 4.58% is reported in Tapioca (table 1). But according to my practical observations maximum post-harvest losses occurs in most of the leafy vegetables because of high water content which starts evaporates immediately after harvesting and causes physiological weight loss which results looseness and drying of leafy vegetables. In fruit crops, Guava and Papaya have maximum post-harvest losses due to weight loss and enhanced ripening which results drying and shriveling of fruit skin and deterioration in qualities and taste within 1 or 2 days.

Table 1: Harvest and Post-harvest Losses of Major Horticultural Crops

S. No.	Name of Crops	% Average Loss (CIPHET Nanda *et. al.*, 2012)	% Average Loss (CIPHET Jha *et. al.*, 2015)
1.	Apple	12.3	10.39
2.	Banana	6.6	7.76
3.	Citrus	6.4	9.69
4.	Grapes	8.3	8.63
5.	Guava	18.01	15.88
6.	Mango	12.7	9.16
7.	Papaya	7.36	6.70
8.	Sapota	5.8	9.73
9.	Onion	7.5	8.20
10.	Tomato	13.0	12.44
11.	Cabbage	6.9	9.37
12.	Cauliflower	6.9	9.56
13.	Green Pea	10.3	7.45
14.	Potato	9.0	7.32
15.	Tapioca	9.8	4.58

Source: CIPHET Report

Causes of Post-Harvest Losses

Horticultural crops particularly fruits and vegetables have high moisture content, tender texture and high perishability. If they are not handled properly, deterioration of valuable nutrients will start and produce may rot within few hours or day. There several causes of post-harvest losses which can be divided into different categories.

1. **Metabolic:** All fresh horticultural crops are live organs. The natural process of respiration involves the breakdown of food reserves and the aging of these organs.
2. **Mechanical:** Due to their tender texture and high moisture content, fresh fruits and vegetables are very susceptible to mechanical injury. Poor handling, unsuitable containers, improper packaging and transportation can easily cause bruising, cutting, breaking, wounding and other forms of injury.
3. **Developmental:** These include sprouting, rooting, seed germination, which lead to deterioration in quality and nutritional value.
4. **Parasitic diseases:** High post-harvest losses are caused by the attack or infection of fungi, bacteria, insects and other organisms. Micro-organisms attack fresh produce easily and spread quickly, because the produce has plenty of nutrients and moisture to support microbial growth.
5. **Physiological deterioration:** Fruits and vegetable cells are still alive after harvesting and continue their physiological activity and physiological deterioration occurs by enzymatic action which leads over-ripening, senescence and aging. Physiological disorders may occur due to mineral deficiency, low or high temperature injury or undesirable atmospheric conditions, such as high humidity.
6. **Lack of market demand:** Due to high production there is glut in the market during harvesting season which cannot be sold and consumed at a time. If there is low demand, lack of infrastructures for storage and marketing facilities and informations may lead great amount of post-harvest losses in fruits and vegetables. This situation occurs most frequently in areas where transportation and storage facilities are inadequate. Produce may start rotting in production areas, if farmers are unable to transport it to distant markets or locations.
7. **Consumption:** If consumers are unable to consume surplus produce, losses can occur due to inadequate preservation methods at home, methods of cooking and preparation such as peeling, consumption styles etc.
8. **Others**
   (a) Lack of clear concept of packing house operations.
   (b) Lack of awareness among the growers, contractors and even the policy makers.
   (c) Lack of infrastructure and poor storage facilities
   (d) Late realization of its importance,
   (e) Inadequate technical support.

(f) Wide gap in technologies available.

(g) Inadequate post-harvest quality control techniques.

(h) Unorganized marketing system.

(i) Absence of pre-cooling and cold storage.

(j) Inadequate market facilities and market information service (MIS).

Factors Influencing Postharvest Shelflife of Fruits and Vegetables

According to Chadha and Pal (2015) all factors influencing postharvest shelf life of fruits and vegetables which cause great losses to harvested fruits and vegetables. These factors can be grouped into two major groups (i) biotic factors (ii) abiotic factors.

Biotic factors

i. Stage of maturity
ii. Rate of respiration
iii. Rate of transpiration
iv. Rate of Ethylene production
v. Changes occur in chemical composition
vi. Changes occur in colour and flavor
vii. Tendency of sprouting in some vegetables like potato, onion and garlic.
viii. Pathological infection

Abiotic factors

i. Temperature
ii. Relative humidity
iii. Oxygen balance in storage environment
iv. Light
v. Types of packaging materials
vi. Mode of transport
vii. Mechanical injury or bruishing

Other factors

i. Behaviour of cultivar or species
ii. Harvest maturity
iii. Improper harvesting
iv. Storage conditions
v. Poor farm practices

vi. Postharvest treatments
vii. Poor infrastructures
viii. Ill equipped marketing facilities.

Advantages of Reducing Post-Harvest Losses

i. It helps to extend the period of marketing, prevent stagnation of the produce and crash in prices during peak harvest time.
ii. It helps to regulate supply of high quality produce to the consumers, which ensures higher remuneration to the producers and longer availability of nutritious and healthy produce to the consumers at reasonable price.
iii. The growers will naturally produce more because they can reap high profit for the produce which would have gone to waste otherwise.
iv. Wastage of energy for production and market the produce will be eliminated.
v. It increases the food supply and provides better nutrition to the people with the same amount of non-renewable resources such as land, water, energy, and labour.
vi. It will reduce waste, the problem of garbage disposal and consequent pollution.

Control of Post-Harvest Losses

There are several techniques related to cultural operations, harvesting, post harvest handling and storage of the produce which can control post-harvest losses of fruits and vegetables considerably.

A. Pre-harvest Operations

Proper cultural operations before harvesting sometimes help to prolong the shelf life of fruits and vegetables.

1. **Fertilizer applications:** There are many points to be considered at the time of fertilizer application.
   i. Avoid heavy application of nitrogenous fertilzers
   ii Application of potassic fertilizer is essential because potassium improve the keeping quality of fruits and vegetables by activating or stimulating enzyme activities that are involved in carbohydrates and protein metabolism and is essential for cell organization and structure of cell walls.
2. **Irrigation:** Irrigation should be stopped 15 days before harvesting specially in vegetables like potato, onion, garlic for obtaining tubers/bulbs with maximum storage life. It check sprouting in the storage.

3. **Blanching in cauliflowers:** It is tying of leaves to protect the curd from sun scorching which ensures preferable creamy white of milky colour and good quality and long storability of curds.
4. **Mulching:** Mulching also prolong shelf life of vegetables because mulching materials reduce reflection of soil heat to produce and thereby keep them less hot which checks deterioration.
5. **Spray of Fungicides:** Three pre-harvest sprays of 0.2% Difolitan at an interval of 10 days are effective in controlling post-harvest diseases of tomato,
6. **Pre-harvest spray of inhibitor:** Researches reveal that spray of inhibitors such as 2000-3000 ppm Maleic hydrazide + 0.25% Difolitan spray before 15 days of harvesting reduces sprouting in onion and 500-2500 ppm M H reduces sprouting in potato in storage.

B. Harvesting Techniques

1. Harvesting should always be done at proper stage of maturity.
2. Properly mature harvested fruits and vegetables have longer shelf-life than unmatured harvested one.
3. Harvesting during hot period of the day should be avoided which causes wilting and shriveling of produce.
4. Harvesting should be done carefully without causing any damage to produce.
5. Harvesting should be avoided after rain because conditions are more favourable for rapid multiplication of the micro-organisms responsible for decay.

C. Post-Harvest Operations

1. **Carrying of produces from field or orchard:** Fruits and vegetables should taken and carried carefully from the field or orchard for other operations. Mostly baskets, crates, bags, sacks are used for this purpose.
2. **Washing of produce:** After harvesting, the produce should be washed thoroughly with clean water to remove field heat and micro-organisms from the surface of the produce. Washing should be done either by deeping or spraying the produce.
3. **Trimming:** Rotten, diseased, damaged by insects and discoloured leaves of vegetables should be trimmed off. It will improve the quality of produce
4. **Sorting:** Cut, injured, bruised, shriveled, wilted, decayed, under sized and oversized fruits and vegetables should be sorted out. It will not only improve the quality of produce but also minimize the chances of spoilage and extended the shelf-life of the produce.

5. **Grading:** The harvested fruits and vegetables should be graded according to the shape, size, colour, and maturity. Produce should be free from disease and insect attack which helps to prevent the healthy produce from being containminated by decay causing micro-organisms. Properly graded produce can be stored for longer time than ungraded ones.
6. **Pre-cooling:** Freshly harvested fruits and vegetables are having field heat that means they are hot. High temperature is always detrimental to keeping quality of produce. Thus it is desirable to remove field heat from freshly harvested produce before packing and transportation through the pre-cooling process. It slows down the respiration, reduces water loss, retard ripening and senescence and minimizes susceptibility to micro-organisms. It reduces weight loss and shrinkage thus conserves the weight and extends shelf-life of harvested produce. It is very important when fruits and vegetables are transported to long distance. Pre-cooling in our conditions is generally done by immerging fruits and vegetables in clean water. Water is excellent medium to prevent heat from fruits and vegetables are transferred. This is called hydro-cooling.

   There are various other methods of pre-cooling given as below.

   i. Hydro-cooling
   ii. Air cooling
   iii. Room cooling
   iv. Forced air cooling
   v. Top icing
   vi. Vacuum cooling

   In developed countries pre-cooling is mostly done by keeping the produce in refrigerated trucks and by placing ice in the packages. But under our conditions pre-cooling is generally done by immerging fruits and vegetables in clean water.
7. **Curing:** Curing is the post-harvest treatment of tubers and bulbs crops. Proper curing can considerably reduce spoilage of potato, onion and other bulbs and tuber crops like garlic, sweet potato, cassava, yam etc. in storage. Generally harvested onion and potato is cured in the field or open floor by putting in heap under shade for about two weeks. Research report reveals that curing of onion and potato in perforated plastic crates for 15 day at room temperature is most effective for reduction of losses due to rotting in storage.
8. **Waxing:** Edible wax is used for skin coating. Protective skin coating with edible wax is one of the methods to extend shelf-life of fruits and vegetables at ambient room temperature. Waxing is done by dipping

the fruits and vegetables in solution of wax emulsion (6-12%). It is very helpful in fruits and fruit vegetables like okra, eggplant, bitter gourd, pointed gourd etc.

9. **Post-harvest treatment of chemicals:** Control of decay causing micro-organisms in storage is of vital importance for which some pesticides used are given below.

   i. **Fungicides and Bactricides:** Research reports reveal that pre-harvest spray along with post-harvest application of fungicides and antibiotics significantly prevent decay causing fungi and bacteria.

   Effective fungicides: Benlate, DiathaneM-45, Captan, Benmyl

   Effective antibiotic: Streptomycin (1000 ppm) in spinach and lettuce.

   ii. **Antitranspirant:** Application of antitranspirant like Mobileaf, Kaoline is beneficial in green leafy vegetables. It prevents water loss through transpiration and aid to increase conservation of water in harvested vegetables. Use of antitranspirants that partially close stomata and give additional benefit in moisture conservation especially in leafy vegetables

   iii. **Growth substances:** Post-harvest application of growth substances has been proved to extend the shelf-life of fruits and vegetables. Physico-biochemical processes causing deterioration which are going on inside the harvested fruits and vegetables, that are mediated by indigenous growth hormones principally auxin, gibberellins and cytokinins. Application of Exogenous growth substances can alter different metabolic processes and some of these alterations are beneficial to extend the shelf-life of fruits and vegetables.

   Cytokinenine treatment: Senescence process causes yellowing in many fruits and vegetables especially in cauliflower, broccoli, and cabbage that can be retorted by pre or post-harvest application of Cytokinenine for example B A (N6-benzyldenine a synthetic cytokin at lower dose (5-15 ppm) application is beneficial..

   Auxin growth substances: Auxin growth substances like 2,4-D, 2, 4,5-T have been found effective on delaying leaf abscission of cauliflower when applied as pre or post-harvest application.

   Cycocel: Cycocel (2-chloroethyl-trimethyl ammonium chloride) can prolong the shelf-life of lettuce, broccoli, etc.

   Other chemicals

   Borex: Application of borex (4%) reduces the fruit decay in cucumber when stored at room temperature.

   Calcium nitrate or chloride: Calcium nitrate or chloride in the form of dip extend the shelf-life of tomato and other vegetables by reducing weight loss and checking CO_2 and C_2H_4 production.

All these reports are based on researches.

10. **Packaging**

    Packaging of fruits and vegetables provides protection from external injuries and from being shriveled. Thus, it reduces post-harvest losses considerably and the shelf life of fruits and vegetables is extended. It also provides effective marketing. Thus there are two main functions of packaging.

    (i) To assemble the produce in convenient units for handling

    (ii) To protect the produce during transportation and marketing operations.

    The common packages are Wooden boxes, Bamboo baskets, fibre board boxes, plastic crates, jute bags, Polythene film bags, Paper board boxes lined with polyethylene. Packaging materials like polyethylene film bags, paper board boxes lined with polyethylene can effectively protect the fruits and vegetables from dry air and can prolong their shelf-life. Polyethylene packaging provides the modified atmosphere and also reduces decay and softening of the produce. Handling of the produce during packaging should be prompt to minimize deterioration.

11. **Transport:** Most of the post-harvest losses of fruits and vegetables occur during transportation and distribution. It has been estimated that spoilage of vegetables takes place one third between farm and road head and another one third between road head and consumers. The spoilage of produce during transportation occurs mainly because of decay due to uncontrolled temperature and relative humidity conditions in carriages. Deterioration in transit is enhanced due to inadequate as well as loose packing.

12. **Storage:** Storage is one of the most important aspects of the post-harvest handling of fruits and vegetables crops to reduce or check the deterioration and decays and to extend the period of availability of fresh fruits and vegetables. The produce is stored at suitable conditions of temperature and relative humidity and there are various forms of storage given below.

    (a) **On farm storage:** If the time between harvest and marketing is extended due to poor transport facilities, hardy some fruits and vegetables such as carrot, cabbage, beet, turnip etc can be stored for very short period to check the post-harvest losses to some extent. The crops are generally placed in piles on the ground and covered with straw or jute sheet first and then soil. Sometimes, on farm storage of these vegetables is done in trench or pit.

    (b) **Home storage:** Limited quantity of fruits and vegetables such as ripe pumpkin, sweet potato, yam, well cured potato, onion and garlic can be kept in the shady and cool corner of the house for quite a long time.

(c) **Low cost storage without refrigeration**

i. **Ventilated storage structure for onion and garlic:** This type storage structure is made up of bamboo pipes or wood sticks. Onion and garlic are stored in layers.

ii. **Zero-energy cool chamber:** This is a low cost cool chamber based on the principle of evaporative cooling. The main objective of this is to maintain low temperature and high humidity as compared to ambient condition outside.

iii. **Cold storage:** Cold storage provides low temperature and that is the best method of slowing down metabolic processes and decay of fruits and vegetables and thereby most effectively extends the shelf-life of produce.

5

Processed Products from Fruits and Vegetables: Status and Scope

Processed fruit and vegetable are considered as staple food products throughout the world, and are becoming more popular with the increase of the global urban population. As a result, demand for processed fruit and vegetable products tends to increase with population growth in the world. There is strong demand for processed fruits and vegetables in the developing world over the past few years. In poor countries, people consume less fruits and vegetables due to the relatively high prices. All forms of preserved fruits and vegetables are available only to the urban elite people. But rural mosses producing more than 90 % of these fruits and vegetables are usually deprived of their uses.

Demand of Processed Fruits and Vegetables

Demand of processed fruits and vegetables is increasing in the developing as well as under developing countries of world. The following reasons are responsible for day by day increasing demand of processed fruits and vegetables.

i. **Increasing awareness of importance of fruits and vegetables:** Fruits and vegetables are major source of vitamins, sugar, organic acids, minerals and other phytochemicals including dietary fibers and antioxidants essential for health. In addition they provide variety of colour, shape, taste, aroma and texture to refine sensory pleasure in human diet (Chun Ta Wu, 20020). Due to nutritional superiority of fruits and vegetables there is increasing awareness at consumer level.

ii. **Surplus fruits and vegetables in season:** Considering the increasing demand of fruits and vegetables at consumer level, the surplus fruits and vegetables produced during harvest season can be processed and preserved for future use when there is scarcity of fruits and vegetables during off season.

iii. **Increasing popularity of staple food:** The processed fruits and vegetables are considered as staple food products there is increasing popularity of processed staple food with increase in world urban population. The increasing growth of urban population tends to increase the demand of processed fruits and vegetables throughout the world.

iv. **More demand in developing countries:** It has been reported that demand of processed fruits and vegetables has increased in developing world over past five years.

International Scenario (Trades)*

The demand of food including fruits and vegetables and processed products is increasing with the increasing income of consumers. However, in most cases, food processing infrastructure in these countries is not developed enough to satisfy domestic demand. According to World Bank figures (2014), about 37.0% of the processed products of fruit and vegetables are traded internationally (by value), which is relatively larger to fresh products because of non-perishable nature of processed produce. The leading exporting and importing countries are given below.

(a) Leading Exporting countries:

(i) Preserved fruit and fruit preparation:

China, United States, Germany, Poland Netherlands.

(ii) Fruit and vegetable juice:

Brazil, Belgium, Netherlands, China United States.

(iii) Frozen and preserved vegetables:

Belgium, China, Netherlands France.

(b) Leading importing countries:

United States, Germany, France, Japan and the United Kingdom.

Processed fruits and vegetables: Japan imports a large percentage.

Fruit and vegetable juices: The United States, Germany, the Netherlands, the United Kingdom, France, and Belgium rank among the largest importers.

Frozen vegetables: France, the United States, Germany, Japan and the United Kingdom are the largest importers.

****Source:*** IBISWORLD Industry Report, Global Fruits and Vegetables Processing, August 2014

National Scenario (Trades)

It is reported that India is second largest producer of vegetables and fruits after China with production of fruits 94.38 Million MT and vegetables 182.03 Million MT. But per capita per day gross availability of fruits and vegetables projected during 2017-18 is 201.50 g fruits and 388.72 g vegetables but net availability is only 141 g fruits and 272 g vegetables

(***Source:*** XIIth Five Year Plan, Report of the Working Group on Horticulture and Plantation Crops, Planning Commission, November 2011, Pg. No. 64). The nutritional intake from fruits and vegetables is higher among urban population than that of rural population which results an increasing demand of fruits and vegetables at consumer level because of awareness about their roles in our nutrition and health.

As it has been reported that 20-40% post-harvest losses in fruit and vegetables occurs and that can be minimized by processing of these highly perishable commodities. The processed fruits and vegetables have more shelf life as compared to fresh one and can be supplied throughout the year to fulfill the demand of consumers. Although there are 38608 food processing industries 1133 fruits and vegetable processing industries are registered in the country till 2014-15. Even then it is surprising to note that country is importing fresh and processed fruits and vegetables from different countries to fulfill the demand of the Indian consumers. During the year 2017-18 country has imported worth of Rs. 1255016.61 lakhs fresh fruits and vegetables and 93718.56 lakhs processed fruits and vegetables (Table 1).

Table 1. Import of Fresh and Processed Fruits and Vegetables in India during 2017-18

Qty in MT Value in Rs. Lacs

Fresh			Processed		
Country	**Qty**	**Value**	**Country**	**Qty**	**Value**
U S A	254096.00	561897.50	Afghanistan	5233.70	20396.80
Afghanistan	40087.00	158123.76	China	15670.46	13274.85
Iran	78189.00	101464.79	U S A	6921.34	10797.72
Australia	31595.00	84802.05	Thailand	5592.00	6320.88
Pakistan	151753.00	72684.70	Spain	5097.65	5952.33
China	74332.00	41604.15	Turkey	3193.72	3645.89
Iraq	146002.00	41345.76	United Arab Emirates	3354.05	2712.13
Chile	36427.00	33508.52	Bhutan	4715.37	2612.70
Sri Lanka	12117.00	30749.03	Brazil	1426.09	2226.89
Egypt	44350.00	14886.44	Italy	1646.32	2041.92
Others	141409.00	113949.91	Others	15786.23	23736.45
Total	1010357.00	1255016.61	Total	68636.93	93718.56

Besides import of fresh and processed fruits and vegetables, country is exporting some fresh and processed fruits and vegetables to some foreign countries to earn foreign exchange (Table 2 and 3).

Table 2. Export of fruits and vegetables from India during 2017-18.

Commodity	Quantity (MT)	Value (Rs. Lakh)	Commodity	Quantity (MT)	Value (Rs. Lakh)
Fruits (Fresh/dry)			Vegetables (fresh/chilled)		
Apple	12529.22	3916.59	Cabbage and Lettuce	527.28	84.45
Banana	101314.37	34877.39	Cauliflower and Headed Broccoli	177.86	95.23
Orange	15840.48	3485.36			
Grape	188221.16	189994.86	Onion	1588985.71	308882.23
Guava	1229.75	553.26	Garlic	31,000.22	13,266.07
Litchi	108.24	57.98	Peas	410.11	179.41
Mango	49180.46	38234.02	(Shelled/Unshelled)		
Papaya	9996.37	4150.44	Tomato	47446.09	11421.91
Pineapple	8339.82	3501.44	Chilli	44,000.99	22,074.05
Sapota	1602.60	890.70	Potato (Other than seed)	395748.12	41182.50
Walnut	3124.65	11927.03	Sweet potato	403.73	124.84

The major destinations for Indian fruits and vegetables are UAE, Bangladesh, Malaysia, Netherland, Sri Lanka, Nepal, UK, Saudi Arabia and Qatar. Europe is the main export markets for Indian fruits and vegetables.

Table 3: Major destinations for processed fruits and vegetables

Processed fruits & vegetables	Major exported countries
Dried & preserved vegetables	Sri Lanka, USA, UAE, Germany France, Netherlands
Mango Pulp	UAE, Saudi Arabia, Kuwait, Netherlands, Hong Kong
Pickle & Chutney	UK, USA, UAE, Germany, Canada, Netherlands, Saudi Arabia
Other processed Fruits & vegetables (Tomato paste, jams, juices etc.)	USA, Netherlands, UK, UAE, Indonesia, Philippines, Russia

Source: Agricultural and Processed Food Products Export Development Authority (APEDA).

Challenges and Constraints in the Development: Challenges

There are several challenges faced by agricultural enterprises to feed the increasing population in the world. Besides other challenges, today, one of the main global challenge is how to ensure food security for a world growing population and to ensure long-term sustainable development.

According to the FAO, food production will need to grow by 70% to feed world population which will reach 9 billion by 2050. Further trends like increasing urban population, shift of lifestyle and diet patterns of the rising middle class in emerging economies along with climate change put considerable pressure strain on the planet's resources: declining freshwater resources and biodiversity, loss of fertile land, etc. Consequently, there is a need for an integrated and innovative approach to the global effort of ensuring sustainable food production and consumption (Nellemann et al., 2009; World Economic Forum 2009; FAO/OECD, 2011; Foresight, 2011; EU ERA-NET SUSFOOD 2012-2014).

When we look at the increasing Indian population and urbanization the country is facing the same challenge. Regarding the increasing demand of horticultural produce, the country is importing fresh and processed fruits and vegetables to fulfill the requirement of consumers.

Major Constraints

- Limitations in fruits and vegetables production.
- Lack of awareness on production method, advantages and utilization aspects of nontraditional horticultural crops particularly vegetables
- Lack of appropriate technological
- Inadequate Institutional support, Credit, input supply, training
- No infrastructure such cold storage facilities are available..
- No cold chain and transport facilities.
- Limited research work on these aspects
- Limited skilled persons
- Limited institutional helps
- No fruit and vegetable processing industries.

Scope For Future

Though there are several challenges and limitations in country, the more attentions have been given in the progress of other industries in the country but the progress of fruits and vegetables processing industries is not satisfactory and still need more attentions. More importance should be given to make aware of nutritional importance of fresh as well as processed fruits and vegetables, because production of processed fruits and vegetables will help to solve food and nutrition security problem, malnutrition, prevention and control of chronic diseases and HIV problems. Not only this but fruits and vegetables processing industries will create employments and generate per capita as well as national income by internal and external trades. Therefore fruit and vegetables processing and product development must be given first priority on large or small scale production because there are scopes for this field as given below.

- **Suitable agro-climatic conditions for exploitation of production potential:** All types of suitable tropical, sub-tropical and temperate climatic and soil conditions are available in the country from North to South and from East to West parts of the country to exploit production potential of horticultural crops..
- **Availability of raw materials:** It is fact that India is a second largest producing country in world for fruits and vegetables. This means maximum produce of different fruits and vegetables are available in the country. Due to improper post-harvest management, there are great post harvest losses as mentioned earlier which results less availability of fruits and vegetables to consumers and low consumption rate of these commodities. As it has been mentioned earlier that only 1 percent fruits and vegetables are processed in the world. If emphasis on fruits and vegetables processing units are given the post harvest losses can be minimized and nutritional problems can be solved easily..
- **Availability of cheap human resources:** Because of high population there is no scarcity of human resources and large numbers of laboureres are available at very cheap rate. They can be employed.
- **Marketing of horticultural crops:** There is no proper marketing system for fresh fruits and vegetables in both rural as well as urban areas. But processed fruits and vegetables packings are available only in urban areas specially in establishing Malls and super markets, which are not sufficient.
- **Malnutrition:** This social problem can be solved by increasing the production and availability of processed fruits and vegetables.

Present Needs

1. Exploitation of unutilized fruits and vegetables for processing.
2. Establishment of cottage industries on processing
3. To make awareness about the importance of processed fruits and vegetables.
4. Easy policy for registration
5. Need to widen marketing facilities
6. Training for housewives and interested persons.

7. There is needs to expand horticulture education only for agricultural students because being a teacher I can say though other fields have adapted this course but teacher and students have no clear fundamental of horticulture. As my experience only teachers and students from horticulture field are successful in this discipline.
8. Equipments and tools used for processing should be available at reasonable prices.

6

Food Spoilage: Causes, Effects and Prevention

After harvesting the quality of fruits and vegetables begins to diminish and products is edible for only a limited time then products begin to spoil or rot. Losses occur during harvesting and handling of fruits and vegetables can be minimized by proper techniques but processed food spoilage is bit difficult to control and stop after infection and contamination of microorganism.

Spoilage

Spoilage is deteriorative process which results reduction in quality of food (fruits and vegetables) which become inedible. The changes in food that cause it to lose its desired quality and eventually become inedible is called food spoilage or rotting. After harvesting, the quality of fruits and vegetables begins to diminish and product is edible for only a limited time. The deterioration and spoilage of fruits and vegetables may occur at different ways.

The losses in harvested products begin immediately after detachment of products from their parental plants. After some time, theses products begin to spoil or rot. Spoilage can be defined as the deteriorative process which renders the food inedible or results in reduction in quality. Spoilage of fresh fruits and vegetables usually occurs during storage, transportation and during processing. They also get contaminated with spoilage organisms either from each other or from the containers. Mechanical injuries during harvesting and transportation further aggravates the deterioration process.

Changes in Products During Spoilage

Physical Changes: Changes in colour, flavour, texture and taste etc.

Chemical Changes: Changes in carbohydrate, fats, proteins, vitamins and minerals.

Microbial Changes: Mould, yeasts and bacteria are responsible for many changes.

Types of Spoilage

The spoilage in harvested fruits and vegetables products is mainly of two types.

1. Abiotic spoilage
2. Biotic spoilage

1. Abiotic Spoilage

It is due to different physical and chemical changes in the product i.e. hydrolytic action of enzymes, oxidation of fats, putrification of proteins, browning reaction between proteins and sugars. The physical changes like wilting, cracking, melting are very common. It has less potential for causing health hazardous to consumers. It is of two types physical and physiological spoilage.

(a) **Physical spoilage:** Physical changes are due to dehydration i.e. loss of moisture from the products and degradation of pigments. Generally following physical changes appear in products.

   i. Weight loss due to loss in moisture content

   ii. Shrivelling due to loss of turgidity in cells which occurs due to loss of moisture content.

   iii. Colour fading due to degradation of pigments.

(b) **Physiological spoilage:** This type of spoilage occurs due to physiological aging and changes in chemical constituents and damage of enzymes. Based on these it is also of two types.

   i. **Physiological aging:** It occurs as soon as the biological cycle is broken through harvesting. The process can be prevented and delayed by storing the products in a dry area at as low a temperature as possible..

   ii. **Chemical and enzyme spoilage:** Chemical and Enzyme spoilage occurs especially when vegetables and fruit are damaged by falling or breaking. Such damage can release enzymes that trigger chemical reactions. The same processes can also be triggered by insects. For examples- Tomato become soft, Apple and other fruits become brown. The enzymes can be deactivated by heating the fruit or vegetables. The same effect can be achieved by making the fruit or vegetables sour or by drying them, but the enzymes become active again as soon as the acidity is reduced or water is added.

2. Biotic Spoilage

This type of spoilage is associated mainly with micro-organisms. It may also be caused by insects, animals and rodents. This has more potential for causing health hazards to consumers.

Causes

i. **Microbial causes:** Microorganisms like bacteria, moulds and yeasts are responsible for this spoilage of food, which is known as microbial spoilage.

ii. **Insects and Rodents:** Insects and rodents can also cause a lot of damage to fruits and vegetables and their products not only by eating, but also by passing on micro-organisms through their hair and droppings. The affected parts are especially susceptible to diseases.

Microbial Spoilage

Microbial changes caused by microorganisms such as bacteria, moulds and yeasts. The peel of a fruit or vegetable provides natural protection against micro-organisms. As soon as this shield is damaged by falling, crushing, cutting, peeling or cooking, the chance of spoilage increases considerably. Crushing occurs most often when fruits or vegetables are piled up too high. Microbial spoilage is most important and described in detail.

Types of microbial spoilage: There are two types of microbial spoilage

(a) Spoilage caused by plant pathogen: Plant pathogen attack various parts of the plant used as foods.

(b) Spoilage caused by saprophytes: This type spoilage occurs after spoilage caused by other agents.

Microorganisms causing spoilage: Microorganisms are very small, one-celled animals. There are three types of microorganisms commonly causing spoilage in fruits and vegetables.

i. Bacteria,

ii. Moulds

iii. Yeasts

Bacteria and yeasts cannot be seen with the naked eye, but moulds are often visible because they form visible thin threads (filaments) or a solid cluster.

Bacteria: Various groups of bacteria can attack different fruit and vegetable depending upon their composition such as, lactic acid bacteria, acetic acid bacteria, caliform bacteria and spore forming bacteria. Main groups of bacteria attacking fruits and vegetables are as follows.

(a) Lactic acid bacteria

Lactobacillus plantaru

Lactobacillus bremis

Lactobacillus fermentum

Pediococus sp.

(b) Acetic acid bacteria

Acetobacter aceti

Gluconobacter sp.

(c) Spore forming bacteria

Bacillus subtilis

Bacillus polymixa

(d) Pathogenic bacteria

Salmanella sp.

Shigella sp.

Yeasts: Yeasts are widely found in the environment including the surface of the fruits, leaves and flowers etc. Yeasts degrade starch, pectins and sugars, thus degrades the product by changing the composition. Yeasts are mostly fermentative and produce followings.

i. *Carbondioxide*
ii. *Alcohol*
iii. *Glycerol*
iv. *Acetaldehyde*

Responsible yeast are

i. *Candida* sp.
ii. *Torulopsis* sp.
iii. *Rhodoturola* sp.
iv. *Saccharomyces* sp.

Moulds: Moulds are associated with food products and important fungi and moulds are causing spoilage given below.

i. *Altrrnaria* sp.
ii. *Soergillus* sp.
iii. *Colletotricum* sp.
iv. *Rhizopus* sp.
v. *Sclerotia* sp.

They produce mycotoxin in rotten food which is highly toxic to animals and toxic to human being (Table 1).

Table 1: Mycotoxins produced by moulds in different foods.

S. No.	Mycotoxins	Microorganisms	Food products
1.	Aflafoxin	*Sprgillus flavus*	Barley, Corn
2.	Patulin	*Penecillium* sp	Apples juices
3.	Ochrataxin A	*Spergillus* sp, *Penecillium* sp.	Corn, Wheat
4.	Roquefertine	*Penecillium roqueforti*	Cheese

Negative effects of mycotoxins: When rotten food is eaten that can cause food poisoning and food illness.

(a) **Food Illness:** Any disease caused due to spoiled food when it is eaten.

(b) **Food poisoning:** It is illness caused by consuming the poison present in food.

The poison or toxins are produced by food spoilage microorganisms. Toxin is very potential and can put human life in risk. Food poisoning is caused mainly by two microorganisms.

i) Clostridium botulinum

ii) Staphylococcus aureus

Symptoms

(a) By Botulism (Food poisoning):

i. Fatigue
ii. Digestive disturbance
iii. Dizziness
iv. Headache

Spoiled foods eaten: Canned foods, Beans, Sweet corn.

Control: Poison can be destroyed by heating at 80 °C for 30 minutes.

(b) By Staphylococcus (food poisoning)

i. Salivation
ii Vomiting

Foods eaten: Meat, Meat products, Fish products, Milk

Control: It can be destroyed by heating at 121.1°C. for 10 minutes.

Many of the moulds secrete toxic compounds like aflatoxins, patulin etc.

Factors affecting growth of microorganisms: Like humans, micro-organisms require certain minimum living conditions

1. Sufficient water
2. Oxygen

3. The right degree of acidity
4. Nutrients
5. The right temperature

Water: Water is necessary for maintaining many physical processes. But high moisture content always favour the rapid growth of microorganisms. Where there is a shortage or lack of water micro-organisms cannot grow, such as in dried legumes, therefore drying is one way to prevent spoilage. It can also be prevented by adding salt. Salying absorbs some amount of water and remaining water becomes unsuitable for microorganisms survival. The same effect can be achieved by adding sugar to fruit. Enzymatic spoilage is also inhibited by drying.

Oxygen: Most micro-organisms need oxygen. If there is a shortage of oxygen, it is difficult for bacteria to survive

Acidity: Bacteria grow best in an environment that is not too acidic. Less acidic products are therefore especially susceptible to bacterial spoilage. Adding acidity to products slows down the process of microbial spoilage.

Temperature: For survival micro-organisms need a temperature of between 5 and 65°C. When temperature is above 65°C, it becomes very difficult for them to survive. They can be destroyed by (i) Boiling or (ii) Heating

i. **Boiling:** They definitely die if boiled, as long as they are boiled for a certain length of time, such as 10 minutes.
ii. **Heating:** When heated, the micro-organisms slowly die off, but not all at the same time. Heating at temperatures lower than 100°C thus has to be sustained for a longer period.
iii. **Storing at low temperature:** The growth of micro-organisms is also slowed down significantly at temperatures between 0 and 5°C (as in a refrigerator), which makes it possible to store the food products for a few additional days. At temperatures below 0°C microbial growth is stopped completely, but the micro-organisms themselves remain alive.

Do micro-organisms grow differently on vegetables and fruit?

- The spoilage in vegetables and fruit occurs by different microorganisms.
- Fruits, which are usually somewhat acidic, are very susceptible to the growth of yeasts and moulds.
- Vegetables are generally less acidic, and their spoilage is usually caused by bacteria.

Minimizing Negative Effects of Micro-Organisms In Food

Eating rotten food can cause contamination or poisoning, to prevent them from reaching our food. Therefore it is important to work as hygienically as possible when handling fruits and vegetables,

The following practices are therefore recommended:

1. Wash your hands thoroughly with hot water and soap before beginning to prepare food.
2. Make sure that kitchen utensils and appliances are well cleaned and disinfected.
3. Always store food in a clean place.
4. Use herbs and spices as little as possible, because they are an important source of contamination.
5. Use clean and pure salt only, if the salt is not pure, heat it on a dry, metal sheet above the fire.
6. Allow only clean drinking water to come in contact with fruits and vegetables.
7. Never allow anyone who is sick or has open wounds to come in contact with food that is to be preserved.

Source: Consulted reference:

(i) Agrodok 3:Preservation of fruit and vegetables by Ife Fitz James and Bas Kuipers © Agromisa Foundation, Wageningen, 2003. Fourth, completely revised edition: 2003.

(ii) Postharvest Management and Processing of Fruits and Vegetables–Instant Notes. by Satish Kumar Sharma (2010). New India Publishing Agency, Pitam Pura, New Delhi-110088.

7

Principles of Processing and Preservation of Fruits and Vegetables

The fruits and vegetables are highly perishable in nature. The losses in harvested fruits and vegetables start immediately after harvesting and the resulting spoilage caused by biochemical changes (enzymatic) and activity of microorganisms such as bacteria, moulds and yeast. The losses caused by the microorganism in fruits and vegetables products can be prevented by controlling or destroying these organisms in order to preserve fruits and vegetables by processing and preservation means. The processing and preservation is one of the most important activities among various types of activities of Horticultural Science.

Processing: The processing can be defined as an activity, which is performed to maintain or improve the quality or to change the form or characteristics of the product. Processing operations are undertaken to add value to horticultural produce after their production. The main purpose of processing is to minimize the qualitative and quantitative deterioration of produce after harvest.

Preservation: Preservation can be defined as the science which deals with the methods of prevention of decay or spoilage of produce and allow it to be stored in a fit condition for future use.

Fruits and vegetables can be preserved for short and long term. For short term preservation can be achieved by proper handling of produce and low temperature storage. Long term preservation can be achieved by canning, drying and dehydration, fermentation and by using chemical preservatives.

Objectives of Processing and Preservation

(i) A good method of preservation is one that slows down or prevents altogether the action of the agents of spoilage.

(ii) During the process of preservation, the fruits and vegetables should not be damaged. To achieve this, certain basic methods are applied on different types of fruits and vegetables. For example in early days, very cold weather condition or ice was used to prevent spoilage. Thus, very low temperature became an efficient method for preventing spoilage of fruits and vegetables.

Principles of Processing and Preservation

There are three major principles of processing and preservation of fruits and vegetables which are considered important in decreasing microbial and self decomposition of fruits and vegetables on which various methods of processing and preservation are based on destroying microbial activities and inactivation of enzymes. Thus creating unfavourable conditions which are not suitable for microorganisms survival and that also lower enzyme activities or inactivate the enzymes found in produce. These enzymes can be inactivated by changing their conditions such as temperature and moisture. For example when peas are preserved first it is put boiling water for few minutes. It is one of the methods of preservations known as blanching for inactivates enzymes and thus, helps in preserving the produce.

1. **Prevention or delay of microbial decomposition:** The microbial decomposition of fruits and vegetables can be prevented or delayed by following activities.
   (a) **Keeping out the microorganisms:** This can be achieved by following techniques.
      i. **Asepsis:** It is prevention of entry of microorganisms into fruits and vegetables by not opening the natural covering or skin which is a protecting layer of fruits and vegetables.
      ii. **Proper packaging:** The entry of microorganisms can also be prevented by proper packaging of fruits and vegetable products which will not allow the entry of microorganisms.
   (b) **Removal of some microorganisms or reducing microbial loads:** The intensity of spoilage caused by microorganisms can be reduced or minimize by removal of some microorganisms i. e. reducing microbial load from the fruits and vegetables products. This can be achieved by following operations.
      i. **Washing:** Washing with fresh water removes adhering microorganisms on the surface of fruits and vegetables.
      ii. **Trimming:** Trimming is the removal of damaged or spoiled parts of the fruits and vegetables by microorganisms.
      iii. **Filtration:** Juices of fruits and vegetables can be filtered through membrane to remove microorganisms.
      iv. **Centrifrifugation:** It is also applied in juice to remove microorganisms.
   (c) **Hindering growth and activity of microorganisms:** This principle deals that how to check the growth, activities and multiplication of

microorganisms. This can be achieved by creating unfavourable conditions by the techniques given below.

i. **Moisture removal (low moisture content):** Fruits and vegetables contain 70-95% water and high moisture content favour rapid growth and multiplication of microorganisms due to which spoilage intensity is more or fast. Removal of water below the required level for growth of microorganisms and other activities stops microorganisms activities and also reduces the rate of enzymatic activity and biochemical reactions. The moisture can be removed or minimized to stop the activities of micro-organisms and enzymes by considering the following principles of processing and preservation.

If the moisture content of the product is removed to below the suitable level of moisture for growth of microorganisms that inactivate the microorganisms. The moisture content of products can be removed or minimize by the following procedures.

Drying and Dehydration: The terms drying and dehydration means the removal of water.

Drying: It is the removal of moisture from fruits and vegetable products by using non-conventional energy sources like sun and wind. It is done under open condition to bring unfavourable level of moisture content of fruits and vegetables. For example: Sun drying and Solar drying.

Dehydration: It is the process of removal of moisture content of fruits and vegetable products by application of artificial heat under controlled conditions of temperature, humidity and air flow. In this process a single layer of fruits, whole or cut into pieces or slices are spread on trays which are placed inside the dehydrator. The initial temperature of the dehydrator is usually 43°C which is gradually increased to 66-71°C for fruits.

Vacuum drying or Vacuum evaporation of moisture: This is the removal of major part of water content of products by vacuum evaporation or fractional freezing. In vacuum drying, the temperature of the produce and the rate of water removal are controlled by regulating the degree of vacuum and intensity of heat input

Freeze drying: In this method, the produce is dried by sublimation process, i.e., just converting the produce into ice without passing through the liquid form of water by means of vacuum plus heat applied in the drying chamber. In this method, product is frozen first and then water is removed by vacuum and application of heat which occurs simultaneously in same chamber.

In this process produce is first frozen at -18 °C on trays in the lower chamber of a freeze drier and then frozen material dried (initially at 30 °C for 24 hours and then at 20 °C) under high vacuum in the upper chamber. The product is highly hygroscopic, excellent in taste and flavour. For example Mango pulp, orange juice, passion fruit juice and guava pulp are dehydrated by this method.

ii. Acidity control (Low pH): Under acidic condition microbial growth and activities are checked or stopped which help in preservation of produce. This conditions can be achieved by the following techniques.

   Fermentation: Fermentation is the decomposition of carbohydrate (CHO) by enzymes. Fermentation of produce results in the production of organic acids and alcohol which create unfavourable conditions for survival of microorganisms. The production of organic acids and alcohol not only help in preserving the produce but may also produce distinctive new food products

   Acidic addition: Many processed foods and beverages need the addition of acids to impart their characteristic flavour and taste in the final product because acid provides desired flavour and taste. They adjust the sugar and acid ratio in the food and proper balance flavour of the food. They also play role in controlling the pectin-gel formation. The acidic condition is unfavourable for growth and activities of microbes. Fruits are acidic in nature and not spoiled more as compared to vegetables which are alkaline in nature. Main acids used are:

   Acetic acid (Vinegar): Acetic acid is commonly used for pickles, chutney, sauce and ketchup, just to inhibit the growth of micro organisms

   Citric acid (Lime juice): Citric acid is used for preparation of jam, jelly, squash, nectar etc. just to increase the acidity

   Lactic acid (Lactose): Lactic acid is used for the formation of curd from milk, raw flavour, and specific to pickles

iii. **Anaerobic conditions:** It is the reducing the availability of oxygen in the air by addition of carbon dioxide and suphur dioxide in air. When KMS is added in fruits and vegetable products, sulphur di oxide is released and helps in preservation of products. This not only retards the growth of microorganisms but also prevent oxidative changes in nutritional composition of products.

iv. **Osmotic or hypertonic solution:** This condition is created by addition of sugar and salt.

   Sugar concentration: Sugar concentration above 70% in fruit and vegetables products prevents its spoilage by microorganisms. For example Jam, Jelly, Preserves, Candies.

Salt concentration: Salt concentration about 15-20% is sufficient to prevent spoilage of produce.

v. **Spices and Oils:** Addition of spices and oils help to preserve products.

Spices: Spices have been added in food products as a flavouring agent for thousands of years. These compounds which are added to foods are termed as food additives

1. Spices are plant products which are used in flavouring the foods and beverages to enhance the flavour, colour and palatability.
2. They act as antibacterial and antifungal agents.
3. They impart colour as agent.

Oils: The principle behind preservation in oil is air cannot pass through. Thus exhaustion of air as in the case of canning works.

vi. **Presence of High Alcohol level in food products (Alcoholic fermentation.):** Growth, activities and multiplication of microorganisms can be checked by presence of high alcohol level in fruit products. For example fermented beverages contain high level of alcohol varies 5-42% which prevent spoilage.

Wine: Wine is prepared by fermentation of grape juice which contains alcohol about 11-14% but may be as low as 7%.

Frementation is decomposition of carbohydrates by yeast especially yeast species *Saccharomyces cervicea.*

vii. **Presence (Use) of preservatives:** Preservatives are substances or chemicals that are added to products such as fruit beverages and others to prevent decomposition by microbial activities or by undesirable chemical changes. They are added in very low quantities (0.2%) which do not alter organolaptic and physicochemical properties of products or only very little. In general, preservation is implemented in two modes, chemical and physical. Chemical preservation is adding chemical compounds to the product. Physical preservation is processes such as refrigeration or drying.

Preservatives: They reduce the risk of food borne infections, decrease microbial spoilage, and preserve nutritional quality. Preservatives are substances which are added to products just to retard, inhibit or arrest the activity of micro organisms.

Basis of Adding Preservatives

Preservation of food products containing chemical food preservatives is usually based on the combined or synergistic activity of several additives, intrinsic (internal) product parameters (e.g., composition, acidity, water activity) and extrinsic (external) factors (e.g., processing temperature, storage atmosphere and temperature). This approach minimises undesirable changes in product properties and reduces concentration of additives and extent of processing treatments. Chemical food preservatives are applied to foods as direct additives during processing, or develop by themselves during processes such as fermentation, salting, curing, fermentation and smoking.

Type of Preservatives

All the preservatives used in preservation of fruits and vegetable products can be categorized into two groups.

i. **Natural preservatives:** Sugar, commercial salt, and acetic acid.

Salt and Sugar: Salt and sugar bind with water molecules and thus act as a dehydrating agent in foods and impair the conditions under which pathogens cannot survive.

Sugar: The concentration of 68-70% is used for preparation of jam, jelly, marmalades etc. sugar act as a preservative by osmosis and not as a true poison for microorganisms. It absorbs most of the available water, so little water available for the growth of microorganisms.

Salt: The concentration of salt 15-20% is used for the preparation such as pickles. Salt inhibits enzymatic browning and discolouration and also acts as an antioxidants. Curing with salt is used with certain fruits and vegetables products like sauerkraut and pickles.

Salt exerts its preservative action by:

- Causing high osmotic pressure resulting in the plasmolysis of microbial cells.
-Dehydrating produce and micro organisms by tying up the moisture.
- Ionizing to yield the chloride ion which is harmful to micro organisms, and
- Reducing the solubility of oxygen in water, sensitizing the cells against CO_2

Spices: Spices are plant products which are used in flavouring the foods and beverages to enhance the food flavour, colour and palatability. They acts as antibacterial and antifungal activity. They impart as colour agent. Beside these spices also have antibacterial and antifungal activity and act as preservative in pickles

Oils: Oils prevent air to pass through in products.

Vinegar: Vinegar is the acidic element which creates acidic environment under which microorganisms especially bacteria cannot thrive. It is formed by organic process when wine or any other fruit based or grain based alcoholic brew is exposed to a bacterial reaction from the alcohol into acetic acid. Vinegar is prepared from rice or fruits. It is used in small quantity to add flavor and moisture to otherwise bland or strong casting ingredients.

Acetic acid: Acetic acid is commonly used for pickles, chutney, sauce and ketchup, just to inhibit the growth of microorganisms

Citric acid: Natural sources of citric acid are some fruits like citruses. Most commonly citrus fruit is lime/lemon which is used to add citric acid to products. Citric acid is applied in food products during preparation process through lime or lemon juice.

ii. **Chemical preservatives**

Many permitted chemicals used to kill microorganisms or stop their growth. Chemical food preservatives are those substances which are added in very low quantities (up to 0.2%) and which do not alter the organoleptic and physico-chemical properties of the foods at or only very little. Chemicals that are permitted as food preservatives are listed below.

(a) Commonly used acid food preservatives

1. **Benzoic acid:** Benzoic acid in the form of its sodium salt which is one of the most common chemical food preservative. It is applied through Sodium Benzoate. It is a commonly used as chemical preservative in acid or acidified foods such as fruit juices, syrups, jams and jellies, sauerkraut, pickles, preserves, fruit cocktails, etc. Yeasts are inhibited by benzoate to a greater extent than are moulds and bacteria.

   Sodium benzoate is soluble in water and applied to delay the fermentation in the juices. It is commonly used in the product which are having natural colour such as anthocynin pigment.

   Sodium benzoate is more effective against the yeast.

Rate of application: 750 ppm Sodium benzoate is mostly used in fruit juices, squashes and cordials.

2. **Sorbic acid:** Sorbic acid is generally considered non toxic and metabolized preservative. WHO has set the highest acceptable daily intake (25 mg/kg body weight) for Sorbic acid.

Sorbic acid and its salts are practically tasteless and odourless in foods. Sorbates are used for mould and yeast inhibition in a variety of foods including fruits and vegetables, fruit juices, pickles, sauerkraut, syrups, jellies, jams, preserves, high moisture dehydrated fruits etc.

Rate of application: It is applied at reasonable levels (< 0.3 %).

(b) **Gaseous chemical food preservatives**

1. **Sulphur dioxide:** Sulphur dioxide (SO_2) has been used for many centuries as a fumigant and especially as a wine preservative. It is a colourless, suffocating, pungent-smelling, non-flammable gas and is very soluble in cold water (85 g in 100 ml at 25°C). Sulphur dioxide is dissolve in water, and at low pH levels yield sulphurous acid, bisulphite and sulphite ions. The various sulphite salts contain 50-68% active sulphur dioxide. A pH dependent equilibrium is formed in water and the proportion of SO_2 ions increases with decreasing pH values. At pH values less than SO_2 ions increases with decreasing pH values. At pH values less than 4.0 the antimicrobial activity reaches its maximum.

Sulphur dioxide is used as a gas or in the form of its sulphite, bisulphite and metabisulphite salts which are powders. Metabisulphite is more stable to oxidation than bisulphites, which in turn show greater stability than sulphites. The antimicrobial action of sulphur dioxide against yeasts, moulds and bacteria is selective, with some species being more resistant than others.

The gaseous form is produced either by burning sulphur or by its release from the compressed liquefied form. Sulphur dioxide is used in the preservation of a variety of food products. In addition to wines these include dehydrated/dried fruits and vegetables, fruit juices, acid pickles, syrups, semi-processed fruit products, etc. Besides antimicrobial effects, sulphur dioxide is added to foods for its antioxidant and reducing properties, and to prevent enzymatic and non-enzymatic browning reactions.

Source: SO_2 (Sulphur dioxide) is applied through Potassium metabisulphide (KMS). It is unstable and release the SO_2. It is used for the fruit which have no water solvent pigment (colorless). It cannot be used in naturally coloured juices such as phalsa, jamun because they have the Anthocynin pigment. It cannot also be used in the product which is packed in container because it acts on the tin containers and oil and producs hydrogen sulphide (H2S) which has an unpleasant smell and also form a black compound with the base plate of containers. It is best to control moulds than bacteria.

Rate of application: 350 ppm KMS is mostly used in fruit juice products.

General rules for chemical preservation

1. Chemical food preservatives have to be used only at a dosage level which is needed for a normal preservation and not more.
2. "Reconditioning" of chemical preserved food, e.g., a new addition of preservative in order to stop a microbiological deterioration already occurred is not recommended.
3. The use of chemical preservatives must be strictly limited to those substances which are recognized as being without harmful effects on

human beings' health and are accepted by national and international standards and legislation.

Factors which determine or influence the action of chemical food preservatives

1. Factors related to the chemical preservatives

(a) Chemical composition of preservatives;

(b) Concentration of preservatives

2. Factors related to microorganisms

(a) Microorganisms species: As a general rule it is possible to take the following facts as a basis:

- Sulphur dioxide and its derivatives can be considered as an "universal" preservative; they have an antiseptic action on bacteria as well as on yeasts and moulds;
- Benzoic acid and its derivatives have a preservative action which is stronger against bacteria than on yeasts and moulds;
- Sorbic acid acts on moulds and certain yeast species; in higher dosage levels it acts also on bacteria, except lactic and acetic ones;

(b) The initial number of microorganisms in the treated product determines the efficiency of the chemical preservative.

The efficiency is less if the product has been contaminated because of preliminary careless hygienic treatment or an incipient alteration. Therefore, with a low initial number of microorganisms in the product, the preservative dosage level could be reduced.

3. Specific factors related to the product to be preserved

(a) Product chemical composition;

(b) Influence of the pH value of the product- the efficiency of the majority of chemical preservatives is higher at lower pH values, i.e. when the medium is more acidic.

(c) Physical presentation and size which the product is sliced to- the chemical preservative's dispersion in food has an impact on its absorption and diffusion through cell membranes on microorganisms and this determines the preservation effect.Therefore, the smaller the slicing of the product, the higher the preservative action. Preservative dispersion is slowed down by viscous foods (concentrated fruit juices, etc.)

4. Miscellaneous factors

(a) **Temperature:** Chemical preservative dosage level will be established as a function of product temperature and characteristics of the micro-flora;

(b) **Time:** At preservative dosage levels in employed in industrial practice, the time period needed in order to obtain a "chemical sterilization" is a few weeks for benzoic acid and shorter for sulphurous acid.

Source

i. http://www.fao.org/docrep/V5030e/V5030E0d.htm

ii. https://www.slideshare.net/davidmbwiga1990/lecture-5-chemical-preservation-of-food

iii. agri.org/eagri50/AMBE101/pdf/lec23.pdf)

Some food products in which preservatives are commonly used

i. **Citric acid:** fruit juices; jams; other sugar preserves;

ii. **Acetic acid:** vegetable pickles; other vegetable products;

iii. **Sodium benzoate:** vegetable pickles; preserves; jams; jellies; semi-processed products;

iv. **Potassium sorbate:** fruits; vegetables; pickled products; jams, jellies;

v. **Sulphur dioxide:** fruit juices; dried/dehydrated fruits and vegetables; semi-processed products

vi. **Salt (Sodium chloride):** All fruit and vegetable pickles.

vii. **Sugar:** Sweetened fruit juices, Sugar based products such as jam, jelly, marmalade, Saucses, Ketchup, candy, preserves and sweet pickles. etc.

Killing of microorganisms

The final solution of food preservation is killing of microorganisms by the techniques given below.

i. Application of high heat or high heat treatment

ii. Radiation treatment

iii. High pressure treatment

(i) Applying heat or Heat treatment

Microorganisms load can be reduced by killing them by application of heat in various intensities.

Application of heat also inactivates the enzymes responsible for causing spoilage.

Types of heat treatment

(a) High temperature treatment

i. Pasteurization

ii. Sterlization

iii. Simmering

iv. Blanching
v. Frying

(b) Low temperature treatment

i. Freezing
ii. Slow freezing
iii. Quick freezing
iv. Dehydrofreezing

(i) High temperature (heat) treatment

Blanching: It is slight heat treatment done by using hot water or steam that is applied mostly to vegetables before canning or freezing. Blanching is used before freezing to inactivate enzymes present in products that cause deteriorative reactions to food during frozen storage. Blanching is a primary treatment which have to soften the tissues to facilitate packaging, preserve the original colour and flavor, destroy the certain enzyme which are undesirable, elimination of the air, remove micro-organisms

Pasteurization: Pasteurization is the process of heating a food usually a liquid to or below boiling point for a defined period of time. The purpose is to destroy all pathogens, reduce bacteria, inactivate enzymes and extend the shelf life of food product. This method is commonly used for the preservation of fruit juices at home,. The extracted juice is strained and filled in bottles, leaving sufficient head space for the expansion of the juice during heating. The bottles are then sealed air-tight and pasteurized. It is a mild heat treatment. Fruit juices are pasteurized at such temperature and for such periods as would render them sterile, without impairing their flavour. Usually, the juices are pasteurized at about 85 °C for 25-30 min., according to the nature of the juice and the size of container. Acid fruit juices require lower temperature and less time for pasteurization than the less acid ones. Juices can be pasteurized in two ways.

1. By heating the juice at a low temperature for a High time (LTHT) and
2. By heating the juice at high temperature for a short time only (HTST).

Sterilization: Sterilization is destruction of all viable microorganisms by heat treatment. It also inactivates enzymes by heating.

Canning: Canning is done at or above 100 °C. In case of fruits which are acidic, they are canned at 100 °C, while in case of vegetable those are nonacidic, they are canned at above 100 °C. Here, high temperature can be obtained by using steam pressure; time varies according to the type of foods. Due to anaerobic condition any survivable organism will not grow

Frying: It is the heat treatment in which outer temperature of food is very high generally above 100°C. but internal temperature does not reach 100 °C.

Simmering: It is gentle boiling at about 100°C.

Roasting: It is heating of food materials in which internal temperature reaches about 60-85 °C.

Cooking: It is thermal processing in which food materials are heated or boiled for a specific time at a specific temperature.

ii) Low temperature treatment: Low temperature retards the microbial growth and enzyme reaction because it retards the chemical reactions. This is not a permanent method because some microorganisms can also grow at low temperature.

Cold storage: Storage of products at the temperature lower than ambient temperature.

Refrigeration: In refrigeration, fruits and vegetables can be stored for 2-7 days. Semi-perishable crops, such as potatoes, apples etc. can be stored, in the commercial cold storage with proper ventilation, automatic controlled temperature for one year.

Freezing: It ties up the moisture and increase the concentration of dissolved substances in the food. But sometimes enzymes are active even below the 0°C. In this case before freezing, 'blanching' is necessary for vegetable freezing.

Dehydro-freezing: This is a process where freezing is followed by partial dehydration. In some fruits and vegetable about 50% of the moisture is removed by dehydration prior to freezing.

(iii) Irradiation: The process by which some substance such as a food is exposed to some form of radiation such as gamma rays or X rays. In case of irradiation, the food is exposed to the radiations to kill the survive microorganisms by ionizing and non-ionizing radiation like a, b and g rays. Here food is exposed to electromagnetic or ionizing radiation or various frequencies ranging from low frequency electro magnate to high frequency i.e. gamma rays which destroy the micro-organism present in the food.

2. Prevention and delay of self decomposition of food

The self decomposition of food can be prevented or delayed by following ways.

(i) Destruction or inactivation of food enzymes: This principle deals with the inactivation of enzymes for delaying or reducing spoilage by heating or boiling. Techniques are blanching or boiling.

Blanching: It has been discussed already.

Boiling: It has been observed that when potato and apple are cut and kept for sometimes the cut surface turns brown. When potato is boiled and cut it does not turn brown. This due to application of heat during boiling of potato. Boiling inactivates enzymes (PPO-Polyphenol oxidase) which is responsible for browning in potato and apples.

(ii) Prevention and delay of purely chemical reaction: It has been observed in fruit juice which is kept for long time that turns brown. The changes occur due to chemical reaction occurs in juice which cause juice to turn brown due to non-enzymatic (Maillard) reaction.

Prevention

i. Proper peel causes caging

ii. Use of Antioxidants.

3. Prevention of damaged caused by insects or rodents and mechanical causes.

This principle of fruits and vegetables preservation deals with prevention of damage caused by various external agencies other than microorganisms and enzymes. External agency cause physical damage to fruits and vegetables. Insects and rodent like rats eat peel of fruits in storage, if damaged fruits are eaten any health risk can be experienced. Damaged parts may later on give way for the initiation of microbial and self-decomposition.

Prevention: Proper packaging may prevent damage caused by rodents and insects.

8

Processing and Preservation Methods of Fruits and Vegetables

The basic processing and preservation techniques such as canning, freezing, dehydration, salting, pickling and freeze drying are designed to change perishable foods into more usable form and prevent undesirable changes caused by microorganisms, or by chemical, physical and biochemical reactions in the food itself. Hence an ideal processing technology would involve prevention of growth of microorganisms and at the same time retention of flavor, texture, nutritive value and palatability of the products.

Processing and Preservation

Processing and preservation is the best way of utilizing surplus production of fruits during seasonal gluts.

Major objectives of processing and preservation

i. To change perishable food into more usable forms.
ii. To prevent undesirable changes caused by microorganisms.
iii. To retain flavor, texture, colour, nutritional value and palatability of food products.

Advantages of processing and preservation

i. Helps in converting perishable fruits and vegetables in to durable form.
ii. Fruits and vegetables which are very difficult to eat that can be processed in to a range of highly acceptable fruit product.
iii. Helps in reducing wastage.
iv. Help in making value addition. Value added by changing their form, colour and other such methods to increase the shelf life of the perishable fruits and vegetables.

Processing: The processing can be defined as an activity, which is performed to maintain or improve the quality or to change the form or characteristics of the product. Processing operations are undertaken to add value to agricultural materials after their production.

The main purpose of agricultural processing is to minimize the qualitative and quantitative deterioration of the material after harvest

Types of Processing

i. **Primary Processing:** It is purification of raw materials by removing foreign matters and making the raw materials eligible for processing by grading in different lots or conversion of raw materials into form suitable for secondary processing.

ii. **Secondary Processing:** Processing of primary processed raw materials into products which is suitable for uses or consumption after cooking, roasting, frying etc.

iii. **Tertiary Processing:** Conversion of secondary processed materials into ready to eat form.

Preservation: Preservation can be defined as a science which deals with the methods of prevention of decay or spoilage of food, thus allowing it to be stored in fit conditions for future use. In other words, just controlling the physical, chemical or microbial changes in the foods is called preservation.

Physical Changes: In Colour, flavour, texture and taste etc.

Chemical Changes: In Carbohydrate, fats, proteins, vitamins and minerals.

Microbial Changes: In Mould, yeasts and bacteria below.

These changes can be prevented or controlled by various techniques of processing and preservation described below. Although there are many similarities between the processing of fruit and vegetables, it is important to realize the following differences.

Fruits are nearly all acidic and are commonly called 'high acid' foods. This acidity naturally controls the type of microorganisms that are able to grow in fruit products. The spoilage microorganisms that are likely to be found in such products are moulds and yeasts, which if consumed, rarely cause illness. Processing may be achieved by using preservatives such as sugar, salt and vinegar, and by drying, concentration or fermentation.

Vegetables are less acidic than fruits and for that reason are classified as 'low acid' products. A wide range of microorganisms are able to grow in moist low-acid products, which may lead to spoilage and the possibility of food poisoning. To prevent this, vegetables can be processed by heating to destroy bacteria, or by pickling, salting, or drying to inhibit bacterial growth. Care is needed when processing low acid products, such as vegetables, to minimize the risk of transmitting food poisoning bacteria to consumers.

Methods of Processing and Preservation

There are several methods of processing and preservation that can be grouped into following groups.

1. Physical methods

(a) Removal of water–Dried fruits and vegetables like dried mango slices, Amchur, dried cauliflower and fenugreek etc.

i. Drying

(a) Sun drying

(b) Solar drying

ii. Dehydration

iii. Freeze drying

(b) Irradiation

2.Chemical methods

i. Addition of sugar and heating e.g. Jam Jelly

ii. Salting or Brining e.g., Pickles

iii. Addition of chemical preservatives e. g. Benzoic acid, Sulphur dioxide

iv. Addition of acid e.g., Vinegar

3. Fermentation

(i) Alcohol Production

Different methods of processing and preservation of fruits and vegetables are described separately in proceeding chapters. Here operations done for preparation of raw for processing and preservation are given.

Raw materials Preparation for Processing and Preservation

Before processing of fruits and vegetables some important preparations are required to be done for further processing and preservation. These operations are mentioned below.

1. **Selection:** Healthy, disease free, uniformly ripe or mature fruits and vegetables are taken for processing and preservation.
2. **Washing:** Fruits and vegetables are always washed with fresh clean water to remove dust and dirt adhering on the surface.
3. **Trimming:** Undesirable parts are removed like infected by disease organisms or insects. Part of peduncle or stem attaching with fruit at stem end is also removed.
4. **Peeling:** Peeling is removal of any undesirable part which is followed by washing. It is done in some vegetables and fruits. Peel or skin is removed by peeling knife or with stainless steel knife.

5. **Slicing or cutting pieces:** Fruits and vegetables are sliced or cut into desired shape and size before processing with stainless steel knife.
6. **Blanching:** Some produce requires blanching before processing such freezing or drying. Blanching is the heat treatment given to commodity for inactivating enzymes and killing plant tissues to prevent enzymatic and microbial deterioration. This operation makes raw materials soft. There are several methods of blanching given below.

Method of Blanching

i. Hot water blanching
ii. Steam blanching
iii. Vacuum steam blanching
iv. In can blanching
v. In microwave blanching

Out of all these method, Hot water blanching is commonly followed for processing and preservation of fruits and vegetables. So here we will discuss hot water blanching only.

Hot Water Blanching

Procedure: Hot water blanching is widely adapted and used for this purpose. The hot water blanching is the traditional method of blanching and commonly practiced or used. In this method product is held in hot water (85-100°C) until enzymes are inactivate and then water is cooled and drained out.

Time for blanching: Blanching times for selected commodities use one 8 liters water per kg of produce. Add one minute for each 2000 ft altitude if you live at elevations over 4000 ft.

Table 1: Blanching time in boiling water (minutes)*

Commodity	Time
Broccoli	3
Green Beans	3
Cabbage (wedges)	5
Carrots	5
Cauliflower	3 (add 4 teaspoons of salt)
Collard greens	3
Corn (sweet)	7
Eggplant -	4 (add 1/2 cup lemon juice)
Leafy greens	2
Mushrooms	3 to 5

Commodity	Time
Peas	2
Potatoes (new)	4 to 10
Pumpkin	2 to 3 or until soft
Sweet potatoes	15 to 20 or until soft
Zucchini/summer squash	3

Sources

(i) Chioffi, N. and Mead, S. (1991). Keeping the Harvest. Pownal, Vermont: Storey Publishing.

(ii) McClure, S. (1992). The Harvest Gardener. Pownal, Vermont: Storey Publishing.

(iii) Small Scale Postharvest Handling Practices: A Manual for Horticultural Crops (4th edition). July 2002

7. **Sulphuring:** Some produce needs sulphuring before drying. It is done to preserve natural colour. It also prevents darkening, loss of flavor and loss of Vitamin 'C'. Sulphuring is the final step of pre-treatment. Majority of fruits are treated with sulphur dioxide (SO2) for its antioxidant and preservative effects. Presence of SO2 is very effective in retarding browning of fruits. It also reduces destruction of carotene and ascorbic acid. It is applied by the process called sulphuring.

   Whereas vegetables are steeped in Potassium metabisulpite solution.

Objectives of sulphuring: The main aim of sulphuring is to treat fruits with sulphur dioxide (SO_2)

Procedure of sulphuring: There are two procedures of sulphuring: (i) Exposing fruits in sulphur fumes or smoke (ii) Dipping in Potassium metabisulphite solution.

i **Sulphuring by exposing to sulphur dioxide fumese:** In this sulphuring method fruits are exposed to the fumes of sulphur dioxide produced by burning sulphur. It is done in a sulphur box (Fig. 1) which is closed chamber of non-porous materials (concrete), brick, asbestos sheet or wood capable of holding trays containing prepared fruits.

   Quantity of sulphur: About 100-200 g of sulphur is sufficient for a home drier of seven trays 80 × 60 cm dimension and capable of taking 25 kg of fruit. There is another recommendation burn one tablespoon of sulfur powder per 12g per kg of fruit.

ii. **Dipping in Potassium metabisulphite solution:** In this case sulphurring is done specially in vegetables. Vegetables are dipped in 1% potassium metabisulfite solution for one minute which helps to prevent darkening, loss of flavor and loss of vitamin C.

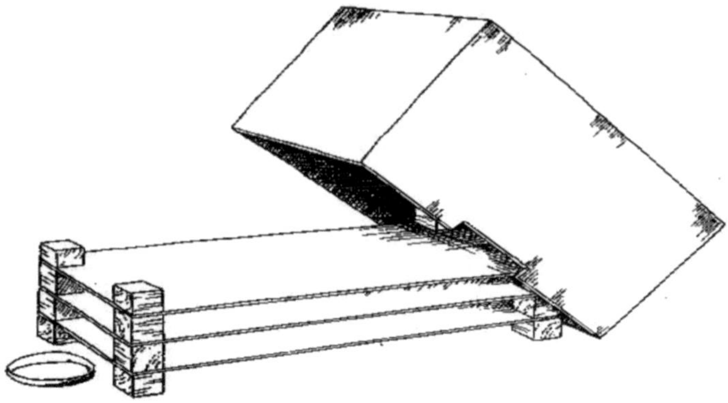

Fig 1 Sulphuring box.*

*__Source:__ Intermediate Technology Publications, 1987. Post-harvest Crop Processing: Some tools for agriculture. Intermediate Technology Publications. London, England

Trays used: Wooden flat trays are generally used for sulphuring and drying

Sulphuring time: Sulphuring is done for ½ hr to 2 hrs depending up on the nature of fruit.

Table 2: Sulfuring times for selected fruits:

Commodity	Sulfuring time	
	Quartered fruits	Halved fruits
Apples	45 minutes	
Apricots	2 hours	3 hours
Cherries	20 minutes 3	0 minutes
Nectarines	2 hours	3 hours
Peaches	2 hours	2 to 3 hours
Pears	2 hours	4 to 5 hours

Sources:

Miller, M. et al. 1981. Drying Foods at Home. University of California, Division of Agricultural Science, Leaflet 2785.

Hobson, P. 1994. Making and using dried foods. Pownal, Vermont: Storey Publishing.

Small Scale Postharvest Handling Practices: A Manual for Horticultural Crops (4th edition). July 2002

Sulfur has been noticed that a source of allergic reactions in some people, so packages of sulfured product should always be clearly labeled. Alternative pre-treatment to prevent browning during drying process is treatment with Vitamin C at the rate of 30 ml ascorbic acid in one liter of lukewarm water. Slice the fruits directly into the solution, remove with slotted spoon, drain well and dry.

9

Physical Methods of Processing and Preservation

In the physical methods of preservation, fruits and vegetables are preserved by (i) adding heat, (ii) removing moisture content of produce to prevent spoilage and (iii) irradiated by Gama rays which kill the microorganisms responsible for spoilage of foods. These methods are described below.

Processing and Preservation by Adding Heat or Heating

In this method, heat treatment is given to destroy the microorganisms which are responsible for spoilage of the produce. Heat treatment is given by many methods.

(a) Pasteurization:
(b) Sterilization
(c) Frying
(d) Canning and
(e) Blanching

Though all these methods have been described earlier, however, they are described in brief.

Pasteurization: It is the heat treatment below 100 °C that kills a part but not all of the microorganisms present in the food.

Sterilization: Sterilization is the heat treatment above the 100 °C to kill all microorganisms.

Frying: It is the heat treatment where outer temperature of produce is very high generally above the 100 °C but internal temperature does not reach above 100 °C.

Blanching: Blanching is the heat treatment given to commodity for inactivating enzymes and killing plant tissues to prevent enzymatic and microbial deterioration.

Methods of Blanching

i. **In hot water:** In this method product is held in hot water at 85-100°C until enzymes are inactivated and then water is cooled and drained. It can be defined in simple way that it is a heat treatment in boiling water at about 100°C for about 3-5 minutes to inactivate the enzymes present in the produce followed by cooling to room temperature. This is a traditional method and commonly practiced.

ii. **In steam:** The concept involves in this system of blanching is to expose product particles to a steam environment for specific period of time based on conveyor spread.

iii. **Steam and water system:** In this system, both steam and water has three stages of heating to achieve maximum control

   First stage: Product is conveyed through steam to provide rapid temperature increase.

   Second stage: Hot water spray provides constant contact between hot water and product to maximize heat transfer.

   Third stage: It is immersion with the conveyor carrying product below water level and ensuring direct heat transfer between individual product particles.

iv. **In microwave:** This system is not commercially used.

v. **Vacuum steam blanching:** It is also not commonly used.

vi. **In can blanching:** In this method raw materials is filled into a can and can lid clinched after adding a small amount of brine (50 ml for 400 g can). The can in upright position is passed on a conveyor over a ribbon burner. The brine is boiled displacing the air in the can with steam and blanching the vegetables. The can is then seamed and thermally processed. This method of blanching is easily adapted to conventional processing lines and will move at normal canning speed.

Removal of Moisture

Moisture content of fruits and vegetables is removed by drying and dehydration methods described as follows.

A. Drying and Dehydration

Most of fruits and vegetables contain 70-95% water (moisture) especially leafy vegetables contain more than 90% moisture and become very tender. High moisture contents favour rapid multiplication and spread of microorganisms. Microorganisms grow only up to certain levels of moisture content in produce and below that level, they cannot thrive well. Hence removal of water or moisture below that level from fruits and vegetables and create low water activity which hinders the growth and activity of microorganisms and fruits and vegetables can be preserved for long periods. This condition can be achieved by drying and dehydration of produce.

Drying: When the removal of moisture or water from the fruit and vegetable is done by application of natural heat known as drying. The best example of application of natural heat is sun drying

Dehydration: When drying is performed under controlled conditions of temperature and humidity is known as dehydration. Thus dehydration is the process of removal of water from fruits and vegetables by application of artificial heat under controlled conditions of temperature, humidity and air flow.

Factor affecting drying and dehydration

i. Composition of raw materials
ii. Size, shape and arrangement of stacking of produce.
iii. Temperature, relative humidity and velocity of air.
iv. Pressure and heat transfer to the surface.

Advantages: There are several advantages of drying and dehydration.

i. **Reduction cost of transport:** The weight of the produce is reduced to ¼ to 1/9 of its original or fresh weight due to drying and dehydration because water from the produce has been lost by evaporation. Thus cost of transport is reduced which depends up on weight and volumes of produce.
ii. **Reduction in storage space:** Before drying produce is bulky because of high moisture content. During drying process moisture content is reduced and bulkiness of produce is reduced. Due to reduction in bulk of produce it requires less storage space.
iii. **No use of preservatives:** Drying and dehydration process do not require any preservative therefore preservatives are not added in product for its preservation.
iv. **Increase in nutrient concentration:** During drying and dehydration there is more loss of water content due to which nutrient concentration increases. Nutrient concentration is very high per unit weight of dried product because of high moisture content nutrients are in more diluted forms.
v. **Low processing cost:** In drying process non conventional energy is used under sun and solar drying and there is not much expenditure on this

process. Therefore cost of processing is low as there is less labour and no sugar required.

Procedure

Three stages are involved.

1. Pre- treatment:
2. Drying of commodity
3. Post-drying treatment

(1) Pre-treatment: Pre-treatments include two operations.

(a) Raw materials preparation

(b) Colour preservation

(a) Raw material preparation

i. **Selection of fruit and vegetable:** Healthy, uniform size and properly mature or ripe fruits and vegetables should be chosen.

ii. **Sorting:** Sorting is done according to size, maturity and soundness of produce.

iii. **Washing:** Washing is done to remove dust, dirt, insect matter, mould spores that might contaminate or affect aroma, colour or flavor of fruits and vegetables.

iv. **Peeling:** Peeling is removal of any undesirable part which is followed by washing. It is done in some vegetables and fruits.

v. **Cutting:** Cutting is done in some fruits and vegetables to appropriate size and form.

vi. **Blanching:** Blanching is the heat treatment given to commodity for inactivating enzymes and killing plant tissues to prevent enzymatic and microbial deterioration.

Method of Blanching: There are several methods of blanching such as (i) Hot water blanching; (ii) Steam blanching; (iii) Vacuum steam blanching ; (iv) In can blanching ; (v) In microwave blanching but hot water blanching is commonly practiced.

Hot water blanching is widely adapted used for this purpose. In hot water blanching is the traditional method of blanching and commonly practiced or used. In this method product is held in hot water (85-100 °C) until enzymes are inactivate and then water is cooled.

(b) Colour preservation

The colour of fruits can be preserve by sulphuring which has been mentioned earlier in detail. This process is done in fruits. Sulphuring is the final step of pre-treatment. It is done to preserve the colour. Majority of fruits are treated with sulphur dioxide (SO_2) for its antioxidant and preservative effects. Presence of SO_2 is very effective in retarding browning of fruits. It also reduces destruction of carotene and ascorbic acid. It is applied by the process called sulphuring, whereas vegetables are steeped in Potassium metabisulpite solution.

In the sulphuring, fruits are exposed to the fumes of sulphur dioxide produced by burning sulphur. Suphuring is done in a sulphur box which is closed chamber of non-porous materials (concrete), brick, asbestos sheet or wood capable of holding trays containing prepared fruits.

About 100-200 g of sulphur is sufficient for a home drier of seven trays 80x60 cm dimension and capable of taking 25 kg of fruit. Sulphuring is done from ½ hr to 2 hrs depending up on the nature of fruit. Generally wooden flat trays are used for sulphuring and drying.

Methods of Drying

For best results when drying fresh produce, fruits should be sliced or quartered, and vegetables should be thinly sliced, chopped or diced. There are various drying methods given below.

(i) Drying by applying Natural heat

(a) Sun drying

(b) Solar drying

(ii) Drying by applying Artificial heat.

(a) Oven drying

(b) Microwave drying

(c) Vacuum drying

(iii) Osmotic drying

Common methods used for removal of moisture from produce are described below.

(A) Sun Drying

In this method produces are directly exposed to sunlight where plenty of sunlight is available for long period. The produce is spread in thin layer on open floor for drying but the main problems in open sun drying are dust, rain and cloudy weather. Therefore drying area should be dust free. Some time produce is covered with a sheet of muslin cloth to avoid contamination of dirt and dust. There is another traditional technique of drying produce in open sun is drying of produce above the ground level. In order to produce dust free clean products materials this method is most suitable and produce should be dried well above the ground level on trays design for the purpose or net fitted on wooden frame. The most drying trays are wooden frames with a fitted base of nylon mosquito netting (Fig. 1).

Sundrying of Spices

Fig.1. Sun drying of some spices on net fitted on wooden frame.

Merits

There is no or less expenditure

Demerits

i. There is chance of contamination of dust.
ii. Air and cloudy weather hinder the drying process.

(B) Solar Drying

It is a modification of sun drying in which the sun rays are collected in side of a specially designed unit with adequate ventilation for removal of moist air. High sun intensity in the unit raises the temperature inside. The temperature in the unit is usually 20-30 °C higher than open sunlight which results in a shorter drying time. The unit is called Sun dryer. Solar drying unit is composed of following parts. Solar drying of fruits will take 2 to 3 days or longer, while most chopped or diced vegetables will dry in 1 to 2 days.

i. Black painted trays
ii. Solar trays
iii. Collectors
iv. Mirror to increase solar energy and accelerate drying.

The black painted tray absorbed more sun rays which results high temperature inside the unit. Drying material is kept on solar tray for drying. Mirror increases solar energy. All these parts help in fast drying processes and reduce drying period. The structure of home made solar dryer is given below (Fig. 2).

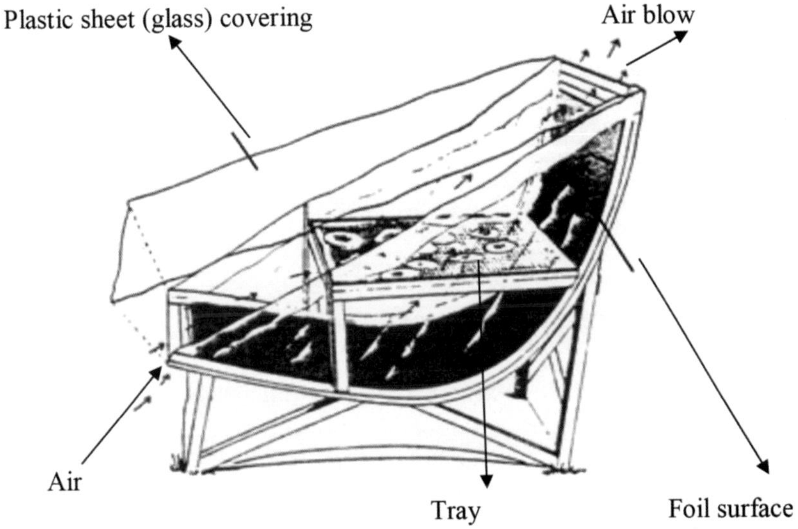

Fig. 2. Home made Solar dryer

Merits

i. Dried products are free from dust and dirt.

ii. Time is saved.

Demerits

There is some expenditure on designing and preparation of unit.

(C) Oven Drying

This method is applied for dehydration of fruits and vegetables. It requires little initial investment

Merits

i. It protects foods from insect and dust.

ii. It does not depends on weather.

iii. Oven dried foods are usually darker, more brittle and less flavorful than dried by dehydrators

Demerits

i. Energy cost tend to be high

ii. It is less efficient

iii. It is difficult to maintain low drying temperature in the oven
iv. Foods are more susceptible to the scorching at the end of drying period. Therefore it is not commonly recommended..

(D) Vacuum Drying

In this method drying is done in vacuum dryers, which is just like tray driers except they are operated under a vacuum and heat transfer is largely by conduction or by radiation. It occurs at low temperature. This method is not commonly used. It includes following parts.

i. Vacuum shelf
ii. Vacuumdrum
iii. Vacuum belt
iv. Freeze driers.

(E) Microwave Drying

Microwaves are portion of electromagnetic spectrum between far infrared and conventional radio frequency region. In this method microwaves are passed through materials like fruits, molecules within the food attempt to align themselves with electric field direction. As they oscillate around their axis heat is produced by intermolecular friction with the product which is responsible for removal of moisture from fruit and vegetables. It is also not commonly used.

(F) Freeze Drying

In this method, the produce is dried by sublimation process, i.e., just converting the food into ice without passing through the liquid form of water by means of vacuum plus heat applied in the drying chamber. In this method, product first frozen then water is removed by vacuum and application of heat which occurs simultaneously in same chamber. In this process food is first frozen at -18 °C on trays in the lower chamber of a freeze drier and then frozen material dried (initially at 30 °C for 24 h ours and then at 20 °C) under high vacuum in the upper chamber. The product is highly hygroscopic, excellent in taste and flavor e. g. Mango pulp, orange juice cone, passion fruit juice and guava pulp are dehydrated by this method.

(G) Osmotic Drying

It is removing of some percentage of moisture from fruits and vegetables by placing them in a concentrated solution of sugar or salt or a combination of both (Fig. 3).

Principles of osmotic drying

Osmosis is the diffusion of water from dilute solution to concentrated solution through a semi-permeable membrane. Fruit cell wall acts as a semi-permeable membrane. Fruits are kept in high concentration of sugar (more than 70%) for 24-48 hours followed by dehydration in hot air.

About 50% of its original weight is reduced by osmotic dehydration process.Diffusion of water through

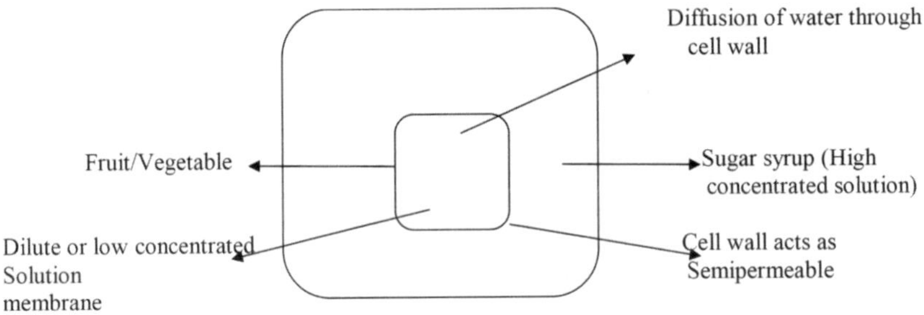

Fig. 3 Osmotic drying

Dehydration

Dehydration is the process of removal of moisture by application of artificial heat under controlled conditions of temperature and humidity and air flow.

Factors affecting dehydration

1. Composition of raw materials
2. Size, shape and arrangement of stacking of produce.
3. Temperature, relative humidity and velocity of air.
4. Pressure and heat transfer to surface.

Flow sheet for drying/dehydration of fruits and vegetables

Fruit and vegetable

↓

Selection
(Mature and free from insects and disease)

↓

Washing

↓

Peeling

↓

Preparation

↓

Blanching

↓

Spreading on flat-wooden trays

↓

Sulphuring
(usually @ 1.8-3.6 kg/ton of fruits)

↓

Drying/Dehydration
(with occasional stirring)

↓

Sweating

↓

Packing
(air tight container or poythene bags)

↓

Storage
(at ambient temperature in dry place)

Products developed by drying and dehydration

A. Dried fruits: Besides temperate fruits like Almond and Walnut, tropical and sub-tropical fruit are also used to prepare dried products.

Mango

i Green fruit dried products: Dried slies or pieces, Dried powder

ii Ripe fruits dried product: Mango leather

Date fruit

Dried date fruits

Date fruit candy

Banana

Banana chips

Grape

Dried Grape- Kismis, Munakka

B. Dried vegetables

Potato - Potato chips,

Cauliflower - Dried cauliflower

Cabbage - Dried Cabbage

Onion and Garlic - Dried onion and garlic

Okra - Dried okra

Cucurbits - Dried bitter gourd and other wild or unexploited cucurbits

Lefy vegetables - Dried leafy vegetables like Fenugreek (Methii) Palak (Spinach beet)

Chilli - Dried chilli

C. Spices

Dried Ginger and Turmeric,

Large and small cardamom etc.

Seed spices- Fennel (almost all seed spices)

Suitability of fruits and vegetables is below given in Table 1 and 2.

Table 1: Suitability of fruits for drying*

Fruit	Suitability for drying	Suitability for Leather	Fruit	Suitability for drying	Suitability for Leather
Apple	Excellent	Excellent	Guava	No	In combination
Apricot	Excellent	Excellent	Melon	Poor	No
Avocado	No	No	Nectarines	Excellent	Excellent
Banana	Good	Fair	Olives	No	No
Papaya	Good	Better	Peaches	Excellent	Excellent
Pears	Excellent	Excellent	Citrus fruits	No	No
Pineapple	Excellent	Excellent	Plum	Good	Good
Coconut	Excellent	No	Pomegranate	No	No
Dates	Excellent	No	Figs	Excellent	No
Grapes	Excellent	Fair	Strawberry	Fair to good	Excellent

Table 2: Suitability of Vegetables for Drying*

Vegetable	Suitability for drying	Vegetable	Suitability for drying
Artichoke	Fair	Okra	Fair to good
Asparagus	Poor to good	Onions	Good to Excellent
Beans green	Fair to Good	Parsley	Good
Beans lima	Fair	Parsnip	Good
Beets	Fair to Good	Peas	Fair to Good
Broccoli	Not recommended	Pepper red	Excellent
Brussels Sprouts	Poor	Chilli	Excellent
Cabbage	Fair	Popcorn	Good
Carrots	Good	Potato	Good
Cauliflower	Poor	Pumpkin	Fair to Good
Celery	Poor	Radish	Not recommended
Collard green	Poor	Rutabagas	Fair to Good
Sweet corn	Good	Spinach	Poor
Cucumbers	Poor	Summer Squaash	Poor to Fair
Eggplant	Poor to Fair	Winter Squash	Not recommended
Galic	Good	Sweet potato	Fair
Horseradish	Good	Swiss chard	Poor
Kale	Poor	Tomato	Fair to Good
Kohlrabi	Fair	Turnip	Fair to Good
Lettuce	Not recommended	Turnip green	Poor
Mushroom	Good	Yams	Fair
Mustard green	Poor	Zucchini	Poor to Fair

**Source:* 32134_27PreservationFVone.9.pdf

(3) Irradiation Methods

Irradiation is the killing of microorganisms by exposing food to various ionizing radiation produced by Cobalt-60 and Cesium 137 under controlled conditions without generation of heat. Therefore it is called "Cold Sterilization". In case of irradiation, the food is exposed to the radiations to kill the survive microorganisms by ionizing and non-ionizing radiation like a, b and g rays. Here, food is exposed to electromagnetic or ionizing radiation or various frequencies ranging from low frequency electro magnate to high frequency i.e. gamma rays which destroy the micro-organism present in the food.

10

Chemical Methods (Use of Preservatives) for Processing and Preservation

In the previous chapter the physical methods of fruit and vegetables processing and preservation have been discussed. The present chapter deals with chemical method of fruit and vegetable preservation. In this method of preservation of fruits and vegetables, chemicals as preservatives are used to prevent the spoilage of fruit and vegetable products. These preservatives react with food (produce) constituents or microorganisms and create unfavourable conditions which hinder growth, activities and multiplication of microbes responsible for spoilage of food products. These chemical agents have maximum permissible limit of their use as law. They can be classified into two classes class I and class II.

Class I: Sugar, Salt and Oils

Class II: Sodium benzoate, Potassium metabisulphite (K M S) and Sorbic acid

(i) **Addition of sugar and heating:** Sugar concentration of 70% or more in fruits and vegetables products prevents spoilage of products. e.g., Jams, Jelly, Preserve, Candy.

(ii) **Salting or Brining:** Salt concentration 15-20 percent is sufficient to preserve food materials e.g., Pickles

(iii) **Addition of chemical preservatives:** Sodium benzoate (Benzoic acid), Potassium metabisulphite (Sulphur dioxide)

Sodium benzoate (Benzoic acid): When sodium benzoate is added in food product as a preservative, it is converted into or released Benzoic acid which interfere with microorganism activities. It is added in acidic fruits at concentration rate of 0.06–0.1%. Benzoate in excess of 1% impart a disagreeable peppery or burning taste to fruit juice.

Potassium metabisulphite (Sulphur dioxide): When KMS is added in food as a preservative it releases Sulphur dioxide (SO_2) in acidic foods which helps in preservation of food materials. The concentration of Sulphur dioxide varies 350-2000 ppm (mg)/kg of product. It is added in fruit juices, fruit pulps, fruit beverages and fruit preserves.

(iv) **Addition of acid (Vinegar):** Vinegar is the acidic element which creates acidic environment under which microorganisms specially bacteria cannot thrive. It is formed by organic process when wine or any other fruit based or grain based alcoholic brew is exposed to a bacterial reaction from the alcohol into acetic acid. Vinegar is prepared from rice or fruits. It is used in small quantity to add flavor and moisture.

Above mentioned methods of fruit and vegetable preservation are applied in value added products development which are described in preparation of different fruit and vegetable preserved products All preservative or chemical gents as preservatives are used for preparation of different value added products of fruits and vegetables.

Value Added Products Development

The following products of fruits and vegetables developed in which preservatives are used

Fruit beverages: Fruit juices, fruit squash, R. T. S., Nector, Cordial.

Sugar based products: Jam, Jelly, Marmalade, Preserve and Candy.

Under the use of preservatives for preparation of different value added products will be discussed separately. In continuation fruit beverages are described here.

Fruit Beverages

Fruit beverages are fruit drinks which are superior to synthetic drinks and differ from them (Table 1).

Table 1: Difference between fruit beverages and synthetic drinks.

S. No.	Fruit beverages	Synthetic drinks
1.	Fruit beverages are prepared partly or fully from fruit juice or pulp with or without additives.	They are artificial and no traces are fruits are added
2.	Fruit drink contains more total carbohydrate and vitamins (A, B and C) and minerals (Iron, calcium etc) and provides more calories.	They contain only water and total carbohydrate only 12%.
3.	Fruit beverages are prepared from fruit juice and pulp	Synthetic or non-fruit beverages do not contain any fruit part for examples Coca cola, Pepsi, Sharbat.

Characteristics of fruit beverages

Fruit beverages have following characteristics.

i. Easily digestible
ii. Highly refreshing
iii. Thirst quenching
iv. Appetizing
v. Nutrionally far superior to synthetic drinks.

Types of fruit beverages

Fruit beverages may be unfermented or fermented. On this basis they are grouped into two groups.

(a) **Unfermented fruit beverages:** Fruit juice which is not undergone alcoholic fermentation is known as unfermented beverages e.g., Fruit juice, Fruit squash Pure fruit juice, Squash, RTS, Nector and Cordial.

(b) **Fermented fruit beverages:** Fruit juice which is under gone alcoholic fermentation by yeast is called fermented beverages e.g., Grape wine, Brandy

Unfermented Fruit Beverages

Unfermented fruit beverages include natural and sweetened beverages such as:

Fruit Juice

Fruit Squash

Fruit R.T.S.

Fruit Nector

Fruit Cordial.

A. Fruit Juice

Fruit juice is a product for direct consumption and obtained by extraction of juice from fruit by pressing process. For example- Mango juice, Apple juice, Orange juice.

Types of fruit juice: Fruit juices are of two types.

(a) Natural fruit juice
(b) Sweetened fruit juice

(a) Unsweetened or Natural fruit juice

This type of juice is extracted from ripe fruits and contains only natural sugars. Natural fruit juice contains 100% fruit juice and natural T. S. S.

Pure fruit juice: This the natural juice pressed out from the fruit and practically its composition remains unaltered during preparation. No addition of sugar, acid, water etc. For example, Orange juice, Apple juice.

(b) Sweetened fruit juice

It is liquid product contains at least 85% fruit juice and 10% T.S. S. Its composition is altered by adding water, sugar, acid, flavor, colours, preservative etc. before consumption. Examples- Squash, Syrup, Cordial, R. T. S.(Table 2).

Table 2: Specification for sweetened fruit juice beverages

Fruit beverage	Fruit Juice (%) Not less than this value	TSS (%) Not less than this value	Acidity (%) Not more than this value	Preservatives (ppm)	
				SO_2	Benzoic acid
Squash	25	40	3.5	350	600
Syrup	25	65	3.5	350	600
Cordial	25	30	3.5	350	600
RTS	10	10	0.3	70	120

Procedure for Preparation of Fruit Juice

1. **Selection of fruits:** Fully ripe fruits having uniform colour are selected.
2. **Washing:** Fruits are washed thoroughly with clean water to remove dirt and dust adhering on the fruit surface.
3. **Sorting:** Partially or completely decayed fruits are removed.
4. **Crushing or Pressing to extract fruit juice:** Fruits of tomato and grapes are crushed to extract the juice but mango fruits are cut in to pieces and pressed continuously to extract the fruit juice. During the extraction, fruit juice should not be exposed to air as oxygen in the air will adversely affect the colour, taste, and aroma and also reduce vitamin content of juice.
5. **Staining:** Extracted fruit juice contains some amount of suspended matter such as broken fruit tissues, seeds, skin etc. these materials are removed through a cloth or sieve. Removal of suspended materials improves the appearance of juice.
6. **Filtration:** Finally suspended particles in the juice are removed with special equipment called Filter Press. Filtration is also done with Woven cloth, Cotton or Wood pulp disc or Porous Porcelain wires.
7. **Preservation of fruit juice and pulp:** Fruit juice after extraction may be spoiled due to fermentation by microorganisms; colour and flavor may change by enzymes and taste and aroma may change due to chemical reactions. Therefore, preservation is done to extend the shelf life of fruit

juice and to prevent decay or spoilage of the juice. Fruit juice can be preserved by the following methods.

(a) Use of high temperature- Pasteurization.

(b) With chemical preservatives- Benzoic acid or Sulphur dioxide (SO_2).

(c) With sugar.

(a) Preservation by Pasteurization

Pasteurization is the process of heating of fruit juice or pulp at about boiling temperature 85°C for 25 to 30 minutes followed by immediate cooling to room temperature to kill microorganisms that cause spoilage of juice.

(b) Preservation With chemical preservatives

Preservatives: According to British Food and Drug Act 1928, preservative is defined as "any substance which is capable of inhibiting, retarding or arresting the process of fermentation, aciditification or other decomposition of food". Commonly used preservatives are Benzoic acid and Sulphur dioxide.

(i) **Benzoic acid:** Benzoic acid is used as preservative through its salt Sodium Benzoate. Sodium benzoate salt is tasteless and odourless. Benzoic acid is soluble in water and more effective against yeast and mould.

Quantity used: The quantity of Benzoic acid depends on the acidity of the products. In case of fruit juice with pH 3.5-4 (acidic) the quantity of Sodium benzoate is 0.06-0.1%. In case low acidic fruit juice like grape juice quantity of Sodium benzoate is 0.3%.

Mode of action: It affects the permeability of cell membrane of the microorganisms. The cellular fluid comes out and microorganisms die.

(ii) **Sulphur dioxide (SO_2):** Sulphur dioxide is used through Potassium Metabisulphite or KMS ($K_2 O_2SO_2$) which is a dry chemical and easier to use than liquid and gaseous Sulphur dioxide. Sulphur dioxide is not used in coloured fruit juice like Grape juice.

Mode action: Two major effects of SO_2

(i) It breaks the S-S bonding Cysteine, Methionin (S containing amino acid) and cause denaturation of proteins so that they are not available to the microorganisms. This process is called Sulphytolysis.

(ii) It checks the supply of carbohydrate to the microorganisms.

(c) Preservation With Sugar

Fruit products containing 66% or more sugar do not generally ferment. If the sugar is present 66% or more in fruit products it will absorb most of the available water which results very little water. So fruit products containing more than 66% sugar have little moisture available to microorganisms so their multiplication is inhibited and they die gradually.

8. **Bottling:** Bottles are thoroughly washed with hot water and drained before filling. A 1.5 -2.5 cm head space is left during filling.
9. **Corking or sealing:** Bottles are sealed with crown corks (by crown corking machine) or with caps (by cap sealing machine).
10. **Labeling and storage:** Bottles are labeled and store in dry and cool place.

Flow Chart for Fruit Juice Preparation

Fruit

↓

Selection

↓

Washing

↓

Sorting

↓

Crushing/Pressing(extraction of juice)

↓

Staining/Filtration

↓

Adding preservatives (Benzoic acid or sugar)

↓

Filling in sterilized bottles

↓

Corking/Capping

↓

Pasteurization

↓

Cooling

↓

Labeling/storage

B. Fruit Squash

This is a type of fruit beverage containing at least the following proportion of ingredients. It is diluted before serving.

Ingredients Proportion

Fruit juice or pulp -25%

TSS -40-50%

Acid - 1.0%

Sulphur dioxide - 350 ppm (not more than this value)

or

Benzoic acid - 600 ppm. (not more than this value)

Mango, orange and pineapple are used for making squash commercially. In general, squash consists ingredients in the proportion fruit juice given in table 3.

Table 3: Quantity of different ingredients required to prepare of different fruit squash.

Squash	Juice (kg)	Sugar (kg)	Water (kg)	Citric acid (g)	SO_2	Benzoic acid (ppm)
Orange	1.0	1.7	1.0	20	350	-
Lime	1.0	1.7	1.0	-	350	-
Pineapple	1.0	1.7	1.0	20	350	-
Mango	1.0	1.7	1.0	20	350	-
Grape	1.0	1.7	1.0	20	-	600

Procedure for Preparation of Fruit Squash.

Almost all steps are taken in the Squash preparation.

1. Selection of fruit
2. Washing
3. Peeling (in case of hard fruit)
4. Cutting of fruit in case of hard fruits
5. Extract fruit juice

6. Mix sugar and citric acid and water
7. Heat till sugar and citric acid are dissolved
8. Cool sugar syrup
9. Filter syrup through a muslin cloth.
10. Blend or mix clean syrup with fruit juice
11. Add preservative either Sodium benzoate or Potassium Metabisulphite (KMS) in a small quantity of boiled and cooled water or cooled squash
12. Add permitted colour and essence in adequate amount
13. Fill the squash in dry sterilized bottles leaving 2.5 cm head space.
14. Close bottles with pilfer proof closer which should be dipped in 1% Potassium meta bisulphate solution.
15. After sealing, wash bottles and dry
16. Label the bottles

Flow Chart for Preparation of Fruit Squash

Fruit

↓

Washing

↓

Cutting or grating

↓

Juice extraction

↓

Staining

↓

Juice measuring

↓

Preparation of sugar syrup
(sugar+water + acid heating just to dissolve)

↓

Staining of syrup

↓

Mixing syrup with juice

↓

Adding preservative
(0.6 g KMS or 1.0 g Sodium benzoate per liter of squash)

↓

Bottling
(Filling of squash in sterilized dried bottles)

↓

Capping

↓

Labeling and storage

Preparation of Orange Squash

For preparation of 2.5 liters of orange squash, following quantity of different ingredients are required.

Ingredients	**Quantity**
Orange juice	2.5 liters
Sugar	1.0 kg
Water	1.0 liter
Lemon juice	3-4 tsp
Potassium Meta bisulphate	¾ ts

Procedure

1. Preparation of sugar syrup: Mix sugar and water and heat to dissolve sugar in water in large sauce pan or pot.
2. Add lemon juice to sugar solution and let it boil.
3. Cool it
4. Filter sugar syrup till it is warm
5. Cool it at room temperature.
6. Extract orange juice

7. Strain the orange juice
8. Add orange juice to sugar syrup and stir
9. Dissolve preservative in small quantity of orange juice
10. Add preservative solution to orange squash and stir it
11. Pour orange squash in sterilized jars or bottles.
12. Close tightly with its lid.
13. Cool for some times at room emperature.
14. Store it in dry and cool place.

Preparation of Lime/Lemon Squash

Ingredients	**Quantity**
Fruit juice	1 liter
Sugar	2 kg (More sugar because lemon juice is very sour)
Water	250 ml
Potassium Meta bisulphate	¾ tsp (350 ppm)

Procedure

Following steps taken during the class by author.

1. Healthy ripe fruits are selected
2. Selected fruits are washed with clean fresh water.
3. They are peeled with help of stainless steel knife.
4. Juice was extracted by Juicer.
5. Juice was measured with help of measuring cylinder.
6. Sugar syrup was prepared by adding sugar + Water in required proportion as given above in the table. No need to add citric acid because Lime/Lemon fruits contain sufficient quantity of citric acid.
7. Mixture of sugar and water was added in juice
8. This mixture was boiled by putting on heater.
9. Boiling mixture was stirred and scams was removed.
10. Preservative was added and mixed properly.
11. Squash was kept for cooling
12. Ready hot squash was filled in well cleaned and sterilized bottles
13. Bottles are kept at room temperature for cooling.
14. Cooled bottles are stored in dry and cool place.

Flow Chart for Preparation of Lemon/Lime Squash

Selection of fruits

↓

Washing

↓

Peeling

↓

Extraction of juice

↓

Sugar syrup preparation (Sugar+Water)

↓

Heat it until sugar dissolved

↓

Mix sugar syrup and fruit juice

↓

Add preservative

↓

Add edible colour if required

↓

Filter mixture

↓

Fill ready squash in bottles

↓

Corking and labeling

↓

Storage

Some photographs shown are taken during class by Instructor

Selected fruits

Peeled Fruits

Peels of Lemn Fruits

Extraction of Lemon Juice

Lemon Juice and Rinds

Measuring Lemon Juice

Mixing Lemon Juice and Measured Sugar

Heating and stirring of Mixture

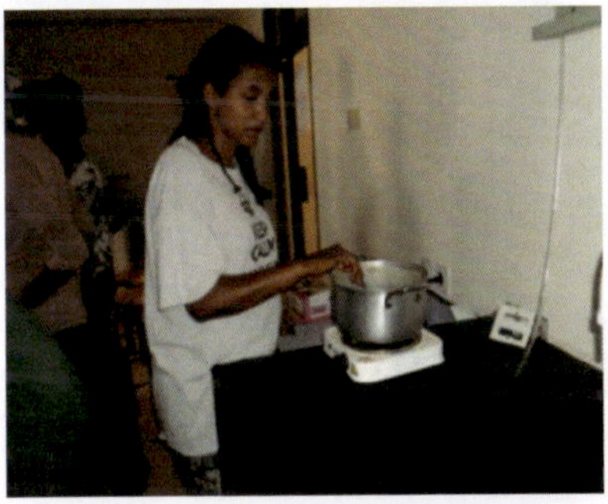

Continuous Stirring of Squash

Boiled ready Lemon Squash

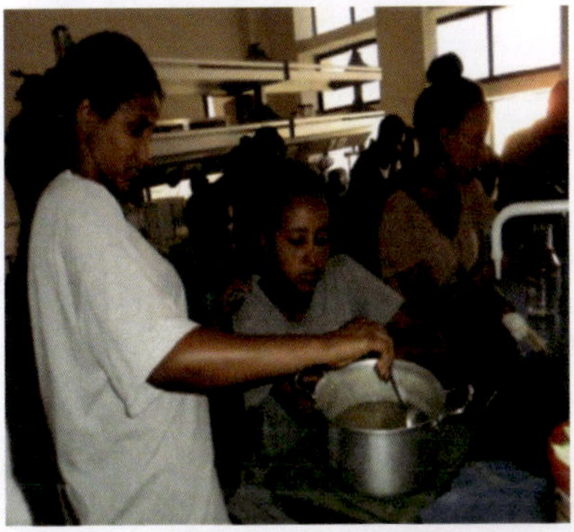

Cooling of Lemon Squash

Filling of Lemon Squash

Lemon squash ready for use

Preparation of Mango Squash

Procedure

For preparation of mango squash followings steps are involved.

1. Select firm but ripe fruit of mango.
2. Wash, peel, de-seed and cut into chunks or pieces.
3. Put pieces in a blender and blend to get thick mango pulp. Sieve the pulp (it will be a little tough to sieve but if some water is added while blending to make the pulp loose and then it is easy to sieve)
4. Measure pure good mango pulp and keep aside
5. Dissolve sugar in water, add the citric acid, boil it and once it is boiled, simmer for some time until it is slightly thick syrup. Switch off fire and allow it cool at room temperature.
6. Pour the strained mango pulp into the syrup.
7. Add lime juice and mix well.
8. Add preservative and mixed well.
9. Transfer the squash into clear jars or bottles and squash is ready for use.
10. To serve, fill 1/4 of a glass with the squash and add water to fill the glass and serve chilled

C. RTS (Ready To Serve)

Ready-to-Serve (RTS) is a type of fruit beverage which contains at least 10 per cent fruit juice and 10 per cent total soluble solids besides about 0.3 per cent acid. It is not diluted before serving, hence it is known as ready-to-serve (RTS). Before preparation of beverages, it is necessary to know the techniques of extraction of pulp/juice from various fruits used for RTS,. The extraction techniques for some fruits have been described earlier. For preparing the beverages the total soluble solids in the pulp/juice and its acidity are first determined and then requisite amounts of sugar and citric acid dissolved in water are added for adjustment of TSS and acidity. On home scale, RTS can be prepared by using the following recipes. It is a beverage which is consumed directly without dilution and contains at least

Fruit juice	10%
T S S	10%
Acid	0.3%

Preparation of RTS (Ready To Serve)

If large quantity of fruit RTS is required such as 20 kg ready RTS following quantity of different ingredients are required.

Table 4: Quantity of ingredients required to prepare 20 kg RTS beverage.

S. No.	Ingredient	Pineapple	Orange	Lime	Grape	Bael	Guava	Mango	Ginger
1.	Fruit juice (kg)	2.5	2.5	1.5	2.75	1.0	1.0.	1.0	0.25
2.	Sugar (kg)	3.0	3.0	3.5	2.5	1.2	1.2	1.2	1.3
3.	Water (kg)	14.5	14.5	15.0	14.7	7.7	7.7	7.7	8.5.
4.	Citric acid (g)	12.5	12.5	-	12.5	28	28	28	30

Procedure

1. Select fruits
2. Extract juice from fruits
3. Mix sugar, citric acid and water for syrup preparation
4. Heat till sugar is dissolved
5. Cool syrup
6. Stir thoroughly after cooling
7. Filter through muslin cloth
8. Mix with juice
9. Add permitted colour and essence on the basis of fruit selected to prepare RTS.
10. To preserve for few days add SO_2 at 70 ppm or Benzoic acid at 120 ppm.
11. Heat it to 90°C
12. Fill it hot in clean and sterilized bottles leaving 2.5 cm head space.
13. Seal with crown cork
14. Hold bottle in horizontal position for 10-15 minutes.
15. Allow them to air cool in vertical position.

Flow Chart for Preparation of RTS

Fruit

↓

Washing

↓

Juice extraction

↓

Filtration of juice

↓

Syrup preparation
(sugar + water heat it till sugar dissolved)

↓

Filtration of syrup

↓

Mixing syrup with juice

↓

Boiling

↓

Adding preservatives, colour, essence

↓

Filling hot in sterilized and dry bottles

↓

Crowning and corking

↓

Heating in boiling water for 15-20 minutes

↓

Cooling

↓

Labeling and storage

11

Preservation by Adding Sugar (Sugar Based Value Added Products)

Sugar also acts as preservative at high concentration 70% or more in food products (fruit and vegetable products). High concentration of sugar in food materials create unfavourable conditions for microbial survival and prevent spoilage. Based on this principle of preservation, sugar based products are prepared from fruits and vegetables so that they can be preserved for long time. This method of preservation with high sugar concentration is principally based on the reduction in moisture content to arrest or stop microbial spoilage. Preparation of sugar based products is one of the most important aspects of home scale as well as industrial level processing of fruits and vegetables. Sugar based value added products include:

i. Jam
ii. Jelly
iii. Marmalade
iv. Preserves
v. Candied

For preparation of these sugar based products quantity or proportion of different ingredients are give as follows (Table 1).

Table 1: Ingredients required for jam, jelly and marmalade.

S. No.	Ingredients	Jam	Jelly	Marmalade
1.	Fruit content (Minimum % in final product)	45	45	45
2.	T. S. S. (Minimum% in final product)	68	65	65
3.	Preservative Sulphur dioxide (ppm) or Benzoic acid (ppm)	40 200	40 200	40 200

Fruit Jam

Fruit and vegetable Jam is a product made by boiling fruit pulp with sufficient sugar to a reasonably thick consistency, firm enough to hold the fruit tissues in position. Apple, pear, sapota (chiku), apricot, loquat, peach, papaya, carrot, plum, straw- berry, raspberry, mango, grapes and muskmelon are used for preparation of jams. It can be prepared from one kind of fruit or from two or more kinds. Proportion of different ingredient for fruit jam preparation is given below.

Ingredients Proportion	
Fruit pulp	45 parts
Sugar	55 parts
TSS of finished product	68%
Acid (Acidity) of finished product	0.5 to 0.6 per cent

Acid may be added according to need. However, fruits contain Citric acid, Malic acid, Trataric acid naturally in sufficient quantity. If acidity is required in less acidic fruits in which sufficient acidity is lacking then addition of acid is necessary as adequate proportion of sugar-pectin-acid is required to give good set of Jam.

Fruit Jelly

Fruit jelly is a semi-solid product prepared by boiling a clear strained juice with pectin containing fruit extract, which is free from pulp after addition of sugar and citric acid. Procedure for preparation of Fruit Jelly is similar as mentioned in the case of Fruit Jam.

Characteristics of perfect jelly

1. It is transparent.
2. It is well set but not too stiff.
3. It should have original flavor of fruit.
4. It should have attractive colour
5. It should not be gummy, sticky or syrupy or having crystallized sugar.
6. The product should be free from dullness.
7. It should not be tough and rubbery.

Fruit jam and fruit jelly seem to be similar but they differ each other (Table 2).

Table 2: Difference between Jam and Jelly:

S. No.	Jam	Jelly
1.	Jam is prepared from fruit pulp.	Jelly is prepared from clear fruit pectin extract.
2.	It is not transparent	It is transparent
3.	It contains minimum T S S 68% as per specification	It contains minimum S S 65% as per specification.
4.	In this flavor and colour are added.	It has original flavor and colour.
5.	The optimum pH is 3.35–3.7	The optimum pH is 3.30

Fruit Marmalde

It is a fruit jelly in which slices of fruit or its peel are also suspended. In the preparation of marmalade, pectin and acid contents are kept higher side than jelly. Bitterness is a desirable characteristic of marmalade

Fruit Preserve

It is a product prepared by keeping a whole mature fruit or its large pieces in sugar syrup till it becomes tender transparent and soluble solid concentration is achieved about 68% or more.

Fruits and Vegetable Candy

It is a product prepared by impregnation of sugar syrup into fruit and vegetable and subsequently drained free syrup followed by surface drying to give a transparent coating of sugar on the surface of fruit pieces. The total sugar content of the impregnated fruit is kept at about 75% to prevent fermentation. The drained syrup can be used for the preparation of next batch of product or in Pickles, Sauce, Vinegar making. Most suitable fruit and vegetable for candy preparation are Pineapple, Anola, Karonda, Papaya, Apple, Peach, Peels of Orange and Ginger.

Procedure for preparation of different sugar based products of fruits and vegetables are described below.

A. Preparation of Fruit Jam

Different steps taken for fruit Jam preparation are:

1. **Selection of fruits:** Fully ripe having good colour, flavor and aroma are selected for preparation of jam. Unripe, immature fruits or over ripe fruits should be not be used for preparation of jam.
2. **Washing of fruits:** Selected fruits are washed thoroughly with clean and fresh water.

3. **Trimming:** Stem portion attached with fruits is removed and bruised and blemishing portion of fruits are also removed.
4. **Preparation of fruits:** There are main three operations done for the preparation of fruits.
   i. **Peeling:** Fruits skin is removed with the help of stainless steel peeling knife.
   ii. **Coring:** Core and stone are removed from the fruits.
   iii. **Cutting of fruits:** Fruits are cut into pieces with stainless steel knife. It is done in the case of hard and tough fruits.
5. **Softening or Blanching:** In the case of hard and tough fruits, fruit pieces are boiled in a small quantity of clean water to soften them.
6. **Addition of sugar and acid:** Required quantity of sugar and acid is added.
   (a) **Addition of sugar:** If the fruit is sour, equal quantity of sugar by weight is added. In the case of sweet fruits only three fourth (3/4 th) by weight sugar is added.
   (b) **Addition of citric acid:** Citric acid is also added at the rate of 1.5-2.0 g/kg fruit.
7. **Mixing sugar and citric acid mixture with fruit pieces:** All ingredients sugar +Citric acid+ fruit pieces are mixed thoroughly and this mixture is allowed to stand for ½ to 1 hr so that sugar is dissolved in the juice released from fruit pieces.
8. **Cooking or Boiling:** The above mixture is cooked slowly on low flame or heat with occasional stirring and crushing till the temperature reaches 105 °C.
9. **Testing end point of jam:** There are two methods to test end point of jam whether jam is ready or need to continue cooking further.
   (a) **Sheet test:** In this method when jam moss is boiling for some time and becomes sufficiently thick in consistency, dip a spoon into the boiling moss and let the product run off the sides of the spoon. If on cooling the product falls off in the form of a sheet instead of flowing readily in a single stream. It means that the end point of product has been reached and product is ready for filling into containers. Otherwise continue boiling till the sheet test is satisfactory.
   (b) **Drop test:** This method is a simplest way to determine the end point of the jam. It is used when no facilities are available. This method is commonly used by housewives. In this method, a small quantity of jam is taken in a spoon during cooking and allowed to air and cooled jam is put into a glass with water. If the drop settles down at the

bottom of the glass without disintegration, it means product is ready. If it disintegrates and splits apart it may need to be cooked more for some time.

10. **Filling or Bottling:** First jars or bottles are kept on insulating material like wooden board or a thick pad of cloth for preventing breakage of glass jars. Hot jam is filled into the jars or bottles.
11. **Capping or Corking:** Screw cap is kept on the bottle head in the inverted position till the jam sets and cools and is closed tightly.
12. **Cooling:** Filled jam bottles are kept in inverted position for 5 minute to sterilize the lid and then allowed to cool.
13. **Labeling:** The jam bottle is labeled with all informations.
14. **Storage:** Jam bottles are kept or stored in a cool and dry place.

Flow Chart for Preparation of Jam

Fruit

↓

Washing

↓

Sorting

↓

Peeling

↓

Cutting fruit into pieces

↓

Softening by heating with some water

↓

Pulping

↓

Removal of stone or core

↓

Preparation of sugar + Citric acid + water mixture

↓

Mixing sugar mixture with fruit pulp

↓

Allow this mixture to stand for 1/3 or 1 hr
so that sugar dissolves fruit juice

↓

Cooking or boiling

↓

Sheet test

↓

Filling or Bottling

↓

Capping

↓

Cooling

↓

Labeling

↓

Storage

Preparation of Mango Jam

The procedure for preparation of jam is explained already as above. The same steps are followed for mango jam preparation. Different steps for mango jam preparation is shown in photographs given below which are taken in the class.

Ingredients	**Quantity**
Fruit pulp	1 kg
Sugar	750 g
Citric acid or Lime juice	1.5 g or 2 tsp.
Water	50 ml

Steps

1. Healthy and uniformly ripen fruits were selected
2. Selected fruits were washed properly with fresh and cleaned water.
3. These fruits were peeled with the help of stainless steel knife.
4. Fruit flesh was removed from stone with the help of stainless steel knife.
5. Required quantity of sugar and citric acid, preservative sodium benzoate were added with fruit flesh
6. Mixture was put on regulated electric heater for boiling.
7. Boiling mixture was stirred continuously for proper boiling.
8. It was boiled till a thick consistency observed in juice mixture.
9. End point of jam was determined by sheet test method.
10. When it is ready it was taken down from the heater.
11. Hot jam was filled in the cleaned sterilized bottles.
12. Lid was kept on mouth of the bottle.
13. Filled bottles were cooled at room temperature and lid was tied
14. Bottles were labeled with a containing all detail of jam.
15. These jam bottles were place in dry and cool place for storage.

The following photographs showing different steps for preparation of Mango jam

Selection of mango fruits

Peeling of mango fruits

Mango Fruit Peels

Peeled mango fruits

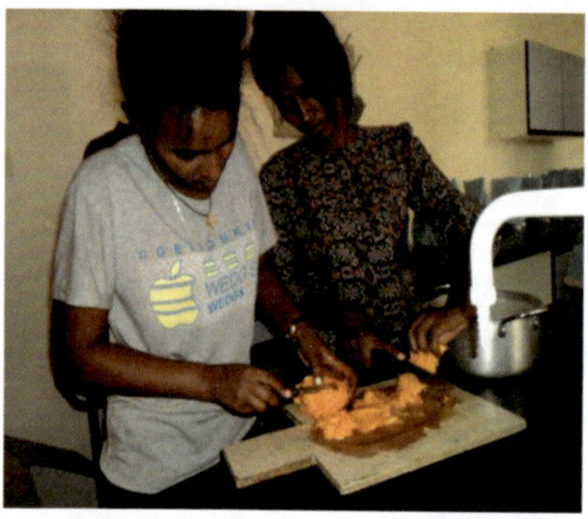

Cutting of peeled mango fruits

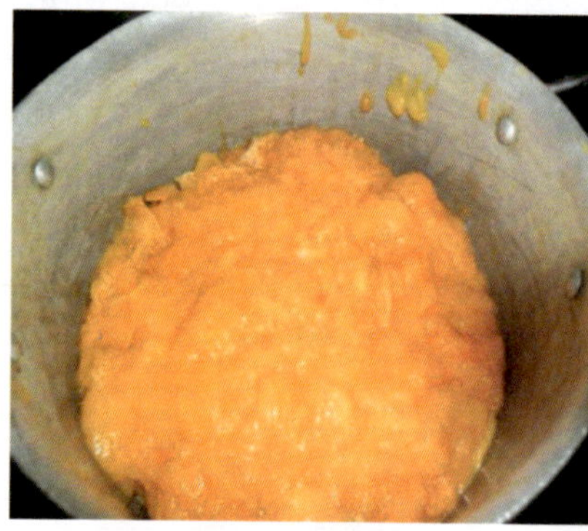

Pulp or Flesh of mango fruits

Mixing of sugar in mango flesh

Squeezing lemon juie in mixture.

Heating mixture of mango

Author showing ready mango jam

B. Preparation of Jelly

i. **Selection of fruit:** Fruit should be rich in pectin and acid content such as Guava, Apple (sour and crab), Plum, Papaya. If the pectin and acid are deficient to make sufficient concentration pectin and citric acid or fruit rich in pectin and acid are added. Both ripe and under ripe fruits can be used but ripe fruits yield good flavor whereas under ripe fruits yield more pectin. Over ripe fruits are avoided as they may disintegrate during boiling and affect the clarity of extract adversely.

   Groups of fruits based on pectin and acid content: Fruits are divided into four groups according to their pectin and acid contents.

   (a) **Fruit rich in pectin and acid:** Sour apple, Lemon, Sour orange, Grapes.

   (b) **Rich in pectin but low in acid content:** Apple, Unripe banana, Pear, Ripe guava.

   (c) **Low in pectin but rich in acid content:** Sour Apricot, Sour peach, Pineapple.

   (d) **Low in pectin and acid content:** Ripe apricot, Peach, Pomegranate, Strawberry and other over ripe fruits.

ii. **Preparation of for fruits**

   (a) Selected fruits are washed thoroughly with clean water

   (b) Fruits are cut into small pieces without peeling e. g., Apple and Guava. In case of Orange, the outer skin (flavedo) is removed to avoid excessive bitterness.

iii. **Extraction of Pectin**

   (a) **Addition of water:** About ½ to equal volume of water is added to fruits like Apple, whereas two to three volumes of water is added to citrus fruits. Water is not added in fruits like grapes and berries.

(b) **Boiling:** After addition of water, fruits are boiled for extraction of juice to obtain maximum yield of juice and pectin. Boiling converts protopectin into pectin and soften fruits tissues. Very juicy fruits do not require addition of water and they crushed only. Boiling should be done in stain less steel containers.

Time for boiling

1. Very juicy fruits 5 minutes
2. Apple 20-25 minutes
3. Grapes and berries 5 -10 minutes
4. Orange 45-60 minutes.
5. Guava 30- 35 minutes.

iv. **Straining and clarification:** The boiled fruit mass is trained through cloth or cheese cloth folded several times. The cloth containing fruit is not squeezed to prevent the solid particles to pass through the cloth. For small scale production, pectin extract is allowed to settle overnight and supernatant liquid can be drained off.

v. **Determination of pectin content of extract:** The pectin content of extract is determined by:

(a) Alcoholic Test: It is rapid test and done by mixing juice with methyl alcohol.

Steps

i. Take 1 tsp pectin extract in a small beaker.
ii. Add 3 tsp methyl alcohol gently along the sides of the beaker.
iii. Carefully rotate the beaker and allow the mixture to stand for a few minutes i.e., 5 minutes.
iv. Observe and interpret as follows on the basis of type of clot or lump formed.

Type of clot formed	Pectin content	Recommended extract and sugar ratio.
Single transparent lump	Rich	1:1
Low firm and fragmented	Moderate	1:0.75
Numerous small granular clots	Poor	1:0.50

(b) **Jelmeter Test:** In this method the pectin extract is allowed to flow through the Jelmeter (on which graduations are marked as ½, ¾, 11/4) exactly for 1 minute. The level to which material has moved indicate the ratio of extract and sugar to be added as in the ratio given below.

(i) 1: ½ (ii) 1: ¾ (iii) 1: 1 (iv) 1: 1¼

v. **Addition of sugar**

Based on pectin test of the fruit extract, quantity of sugar to be added is worked out as follows.

(a) **For extract rich in pectin:** Quantity of sugar equal to quantity of extract is added.

(b) **For extract with moderate pectin content:** 650–750 g of sugar to each kilogram of fruit extract is added.

vi. **Addition of pectin:** For juice rich in pectin jellification occurs without pectin addition. If pectin content is less than 1–2% powder pectin is added to the juice.

vii. **Addition of Acid:** If pectin and acid present in the original fruit extract, acid may be added to the concentrated mass just before to filling into containers.

viii. **Boiling:** Jelly strength increases with the increasing Hydrogen ion concentration until an optimum p H is reached generally 3.2 at 65 % sugar concentration. Jelly strength depends on the quantity of pectin:

(a) Juice is boiled to remove about half of water by evaporation. Remaining water is evaporated until a TSS of 65% is reached

(b) Foam or scum formed is removed during boiling.

(c) Prolong boiling results in the formation of syrup Jelly and darkens its colour.

(d) To limit the time of boiling pectin extract should be boiled first and then sugar is added to it.

(e) For proper inversion, the boiling should be completed within about 20 minutes.

(f) Excessive boiling of jelly results in greater inversion of sugar and destruction of pectin. It should be avoided.

ix. **Judging of end point:** The end point can be judge by sheet test, drop test or refract meter and thermometer.

1. Sheet test

i. Take some boiling jelly in a ladle.

ii. Cool it

iii. Allow to drop it on into pan.

iv. Interpret as

(a) If it drops like syrup, mixture require more cooking.

(b) If it falls like a sheet or flake the end point has reached.

2. Cold plate test (Drop test)

i. Take a drop of boiling jelly or mixture.

ii. Place on a cold inverted steel plate.

iii. Cool it.

iv. Push it slightly with finger.

v. Interpret as: If the jelly is ready, the cooled drop will crinkle and come all at once without breaking.

3. Refractometer Test

In this method T S S of boiling mass is measured by Refractometer and when Refractometer shows T S S 65% Brix, boiling or cooking is stopped.

4. Temperature Test

In temperature test, a solution containing 65% T S S boils at 105 °C. Heating of jelly to this temperature automatically brings the concentration of solids to 65%. The end point of finishing jelly should be 4.5-5.0 °C higher than boiling point of water at that place.

x. **Cooling:** After boiling mixture is cooled slightly up to 85 °C before filling.

xi. **Filling of jelly:** When jelly is ready, it is cooled slightly before pouring into dry and sterilized glass jars or bottles but it should hot. After pouring it is allowed to cool rapidly at room temperature.

xii. **Capping or Corking:** Screw cap is kept on the bottle head in the inverted position till the jam sets and cools. is closed tightly.

xiii. **Labeling and Storage:** The jam bottle is labeled with all informations and bottles are kept or stored in a cool and dry place.

Flow Chart For Preparation of Jelly

Fruit

↓

Washing

↓

Sorting

↓

Cutting fruit into pieces

↓

Boiling in small quantity of water for 20-30 minute

↓

Testing of acidity (Citric acid is added @0.2%)

↓

Straining extract through muslin cloth

↓

Extract

↓

Adding sugar based on pectin content of extract

↓

Cooking the mixture

↓

Removal scum

↓

Judging end point (T S S 56% Brix)

↓

Adding Citric acid and quick mixing

↓

Filling hot into glass jars

↓

Cooling

↓

Waxing

↓

Lidding

↓

Labeling and Storage

C. Preparation Of Marmalade

Marmalade is a fruit jelly but differs in which slices of fruit or its peel are suspended. In the preparation of marmalade pectin and acid contents are kept higher side than jelly. Bitterness is a desirable characteristic of marmalade. Preparation of marmalade is similar to jelly preparation.

D. Preparation Of Preserves

Fruit preserves can be prepared by Apple, Anola, Pineapple, pear, Mango, Papaya, Ash gourd fruits. Preparation of fruit preserve involves following operations.

Primary operations

1. **Selection of fruits:** Fully developed, firm and slightly under ripe fruits are selected. In the case of mango pulp should be yellowish colour.
2. **Washing:** Fruits are washed thoroughly with clean water.
3. **Sorting:** Fruits with juicy pulp are discarded.
4. **Peeling:** Peel is removed by stainless steel knife.
5. **Cutting fruit into pieces:** Fruits are cut into large as well shaped pieces or slices.
6. **Brining:** Fruit slices or pieces are kept in 2-3% brine solution or in water to prevent them from browning or turning black.
7. **Pricking or Puncturing:** It is done by stainless steel fork or bamboo prickers to promote sugar penetration into fruit pieces or slices.
8. **Blanching:** Blanching is done by dipping pieces or slices wrapped in muslin cloth in boiling water for 5-10 minutes. It is done to inactivate enzymes and reduce oxidative discolouration.

Secondary operations

9. **Preparation of preserves:** Preserve is prepared by adding sugar concentrated and packed after addition of preservatives. Preserves can be prepared by two processes.

   (a) Rapid process: In this method a low sugar syrup is prepared. Steps are given below.

   i. Equal weight of sugar to fruit weight is taken
   ii. Water in the ratio of 1:3 (1 part sugar + 3 part water) is mixed with sugar
   iii. Mixture of sugar and water is boiled to prepare sugar syrup.
   iv. Citric acid at rate of 1.5–2.0 g per kilogram of sugar is added during boiling.

v. Fruit slices or pieces are added to the sugar syrup.

vi Boiling is continued with gently heating until the syrup becomes sufficiently thick till the temperature reaches 106°C. Finally sugar concentration in syrup should not be less than 68%.

(b) Slow Process

i. In this process fruits are peeled and blanched

ii. Fruits are cut into pieces or slices.

iii Fruit pieces or slices are kept in alternate layers of equal quantity of sugar as the fruit weight.

iv. This mixture is kept to stand for 24 hrs.

v. The sugar penetrate into the fruit pieces or slices and water exudes out and dissolves the rest sugar.

vi. Second day material is drained over a sieve to separate fruit pieces and sugar syrup.

vii. Syrup is heated to concentrate it up to 60°Brix.

viii. Fruit pieces are put back into concentrated sugar syrup.

ix. Third day same process is repeated and sugar syrup is concentrated to 65°Brix.

x. Fourth day again the same process repeated and sugar syrup concentrated to 70°Brix. Now fruit pieces are ready for packing.

10. **Filling:** Ready fruit pieces are filled into glass jars or bottles and preserved in concentrated sugar syrup of 70°Brix.
11. **Sealing:** Containers are sealed airtight.
12. **Labeling and Storage:** Containers are labeled and stored for few days before consumption.

Flow Chart for Preparation of Preserves

Fruit/vegetables

↓

Washing

↓

Sorting

↓

Peeling

↓

Pricking

↓

Blanching in boiling water with added KMS

↓

Mixing sugar and fruit pieces or slices in alternate layers in the ratio of 1: 1

↓

Keeping this mixture for 24 hrs

↓

Straining over sieve to separate fruit pieces

↓

Sugar syrup concentrated up to 60°Brix by heating

↓

Pouring fruit pieces into sugar syrup

↓

Keep this mixture to stand for 24 hrs

↓

Strain over sieve

↓

Sugar syrup and fruit pieces separated

↓

Concentrate sugar syrup again up to 70°Brix by heating

↓

Put fruit pieces into sugar syrup

↓

Keep for few days

↓

Preserve is ready

↓

Filling in glass jars

↓

Lidding/capping

↓

Labeling and Storage

E. Preparation of Fruit/Vegetable Candy

The process of preparing fruit candied is practically similar to that of fruit preserves. The only difference is that the fruit is impregnated with sugar syrup having a higher percentage of sugar. The total sugar content of the impregnated fruit is kept about 75% to prevent fermentation. The sugar syrup left over from the candy preparation process can be used for preparing another batch of candy of the same kind of fruit

Procedure

1. **Selection of fruit:** Ripe firm fruits are selected.
2. **Washing:** Fruits are washed in clean and fresh water
3. **Peeling and Coring:** Peeling is done remove fruit peel or skin. Core is also removed with coring knife or stainless steel knife.
4. **Cutting:** Fruits are cut into large pieces where necessary.
5. **Pricking:** Fruits and fruit pieces are pricked with stainless steel forks or bamboo pricker.
6. **Blanching:** Fruits are blanched for few minutes in boiling water to make them soft. This assists the absorption of sugar.
7. **Adding of sugar with fruit pieces:** Equal weight of sugar to equal weight of fruit is taken. Then sugar is spread on the blanched fruit pieces in alternate layers.
8. The mixture is allowed to stand for 24 hrs. During this period juice exudes out of the fruit pieces dissolves the sugar
9. **Draining syrup and addition of sugar:** Next day sugar syrup is drained off and enough sugar is added to raise the thickness of syrup up to TSS 60°Brix.

10. **Addition of citric acid:** Citric acid is added at the rate of 1.5-2.0 g per kg of sugar is used at start.
11. **Mixing fruit pieces with sugar syrup:** Fruit pieces are added to sugar syrup.
12. **Boiling mixture:** Sugar syrup with fruit pieces are boiled
13. **Keeping for 24 hrs:** After boiling keep the boiled mixture for 24 hrs.
14. **Continuing the procedure:** The same procedure is followed next day till the TSS raised to 68°Brix. and continued till the TSS reaches to 70°Brix.
15. **For candy syrup concentration:** For candy concentration of sugar syrup is continued till it attains a Brix value of 75-85°Brix or the syrup becomes thick enough to give three or four threads when drawn between two fingers.
16. **Draining of sugar syrup:** Sugar syrup is drained off.
17. **Drying of fruit pieces:** Fruit pieces are dried in the shade for 8-10 hrs or more if pieces are not dried properly. A transparent coating is observed after drying.
18. **Filling ready candy:** The product is transferred into clean dry, wide mouth glass jars
19. **Sealing:** Glass jars are sealed.
20. **Labeling and storage:** Jars are labeled and stored in a cool and dry places.

Preparation of Pumpkin Fruit Candy

Ingredients	**Quantity**
Ripe fruit	1 kg
Sugar	1 kg
Lime juice	5 ml
Water	300 ml (if needed because more fruit juice released out)

Steps

Steps followed in the class while teaching the course.

1. **Selection of fruits:** A ripe and healthy pumpkin fruit is selected
2. **Washing:** Fruit was washed with fresh clean water
2. **Peeling:** Fruit was peeled with a sharp stainless steel knife carefully because skin is hard. there may chance of injuring fingers while peeling.
4. **Cutting:** Peeled fruit was divided into pieces and cut into large cube like pieces.
5. **Forking:** Cube of fruits were forked with stainless steel fork

6. Pumpkin pieces and sugar are put in alternate layers and left for over night.
7. Next day pieces and liquid are separated.
8. **Preparation of Sugar syrup:** Sugar liquid with some additional sugar is added to make seventy five percent sugar by weight of fruit weight was taken and 300 ml of water and 5 ml lime juice per weight of fruit are added in sugar and mixed thoroughly.
9. **Boiling:** The mixture was kept on heater for boiling few minutes so that syrup mixed properly and scam produced is removed.
10. **Mixing fruit cubes with sugar syrup:** Pieces (cubes) are put in syrup gently because syrup is hot and boiled till sugar syrup become bit thick
11. **Testing end point:** Boiling syrup is taken by dipping a stainless steel spoon and cool it. A drop of syrup was taken between finger and thumb and finger and thumb were separated gently. If a thread like structure of the syrup occurs then it is ready and if not then it has to be boiled till thread like structure is not received.
12. **Cooling:** When it is ready it was removed from the heater and stopped boiling and kept for cooling.
13. **Taking out cooked fruit cubes:** Fruit cubes are take out of syrup with the help of fork or spoon and kept on wide plates for drying.
14. **Drying of ready candy:** Cubes are kept for drying for about 48 hrs and covered with muslin cloth to avoid dust.
15. **Changing the sides of cubes:** With help of spoon or fork side of cube was changed for proper drying.
16. **Filling in bottle:** When fruit cubes become dry and shining layer of syrup is formed on the surface of fruit cube, then it is dried. Ready pumpkin candy are filled in properly washed and well dried wide mouth bottles or jars.
17. **Capping:** Lid is put and tied properly
18. **Labeling:** Bottles are labeled mentioning name of fruit candy and other details if needed.
19. **Storage:** Bottles are stored in dry and cool place.

Some photographs taken during preparation of Pumpkin fruit candy during class are shown below.

Director of Lab and a senior teacher of Horticulture with Course Teacher Inspecting ready Pumpkin Candies

Drying of Pumpkin Candies

Pumpkin candies ready for utilization or Storage

12

Preservation with Fermentation Methods (Fermented Products)

Fermentation

The decomposition of carbohydrates by microorganisms and enzymes is known as fermentation. During fermentation process of food organic acids and alcohol etc are produced which act as preservatives and also provide fine flavor and aroma that increase the value of raw materials. Most fermented foods contain a complex mixture of carbohydrates, proteins and fats etc. Fermentation occurs when microorganisms consume susceptible organic substances as a part of their own metabolic processes. Natural fermentation is probably oldest form of food preservation.

Advantages

1. Mild condition of pH and temperature during fermentation maintain nutritional properties and sensory characteristics of food.
2. The flavor or texture of fermented food cannot be achieved by other methods.
3. There is low energy consumption
4. Low capital cost
5. Low operating cost.
6. It is a simple method.

Kinds of Fermentations

During the fermentation process acids and alcohol or a mixture of both acid and alcohol are produced, Therefore based on final products, fermentation process can be classified in to three groups involved in fruit and vegetables preservation..

(a) **Acetic acid fermentation:** It occurs by Acetic acid bacteria. Example is Vinegar. In this Wine, Malts and any other Alcoholic and Sugary products may be converted in to Vinegar (Acetic acid and Alcohol).

(b) **Lactic acid fermentation:** It occurs by Lactic acid bacteria. Lactic acid fermentation occurs in Pickles.

(c) **Alcoholic fermentation:** Alcoholic beverage
The best examples are vinegar, fermented pickles and alcoholic beverages i.e., wine. Their keeping quality depends up on the presence of acetic acid, lactic acid and alcohol respectively.

Thus vinegar, pickles, wine and other fermented products are prepared by fermentation method

Fermentation products from fruits and vegetables are groups as follows (Sasikumar, 2016).

A. Fermented products by acetic acid fermentation: It occurs by acetic acid bacteria.
Any alcoholic and sugary or starchy products may be converted to Vinegar.

B. Fermented products by lactic acid fermentation

(a) By Lactic acid bacteria.
Cucumber dill pickle, sour pickles.

Tomato pickles,

Lemon pickles,

Mango pickles,

Cauliflower pickles.

Kimchi - Mixed vegetables with Chinese cabbage

Tarahana–Vegetables and milk

Lactic acid fermentation occurs as follows

(i) $\text{Carbohydrates} \xrightarrow[\text{Microorganism produce a single product}]{\text{Heterofermentative}} \text{Lactic acid}$

(ii) $\text{Carbohydrates} \xrightarrow[\text{Microorganism produce mixed products}]{\text{Heterofermentative}} \text{Lactic acid + acetic acid} + CO_2 + \text{Ethanol}$

(b) **Lactic acid bacteria with other microorganisms–with Mould.** Example, Soya sauce.

C. Products by alocoholic fermentation: It occurs by yeast, yeast with lactic acid and yeast with acetic acid.
(a) By yeast:
Fruit - wine

Wine–Brandy

Grain mash–Whiskey

(b) Yeast with Lactic acid bacteria

Ginger - Ginger beer

(c) Yeast with Acetic acid:

Cacao bean

Alcoholic Beverages

Alcoholic beverages are prepared after fermentation of sugar by yeast. They contain varying amount of Ethyl alcohol about 5–42%. They are consumed directly or after adding water.

1. **Wine:** Wine is produced by alcoholic fermentation of Grape or Grape juice by yeast (Saccharomyces serevisiae). Alcohol content of wine is 11-14% but it may be as low as 7%.

   Examples:

   Grape wine–Grape juice

   Apple wine–Apple juice

   Honey wine–Honey
2. **Brandy:** It is an alcoholic distillate from fermented grape juice or grape wine.
3. **Rum:** It is distillated from alcoholically fermented Sugarcane juice or Molasses.
4. **Whiskey:** It is distillated from fermented Grain or Wheat.
5. **Beer:** It is prepared by fermentation of Barley grains.

Preparation of Grape Wine

Procedure

1. **Selection of grape fruits or berry:** Well ripe grape fruits are selected
2. **Extraction of juice:** Selected grapes are crushed or pressed to extract the juice.
3. **Filling juice in jars:** Grape juice is filled in jar up to 3/4 th volume.
4. **Addition of sugar:** Sugar is added to achieve 20-25% TSS.
5. **Addition of preservatives:** Sulphur dioxide is added at rate of 50-70 ppm to check the growth of wild yeast and bacteria which may cause undesirable fermentation.
6. **Addition of yeast:** A suitable culture of yeast is added as a starter.
7. **Fermentation process:** Sugar is converted into alcohol by fermentation process. Temperature is maintain at 25-30°C.

8. **Aging of wine:** Wine is allowed to age. During aging process secondary fermentation occurs and the wine loses its raw and harsh flavor which is characterized by good flavour and aroma. Aging takes place about 6-8 months.
9. **Clarification:** During aging period natural clarification occurs. It can also be achieved by fining and filtration.
10. **Pasteurization:** Mature or aged wine is pasteurized at 82-88°C for 1-2 minute to destroy spoilage organisms.
11. **Bottling:** Ready grape wine is filled in bottles.
12. **Labeling:** Packaging and storage: Wine bottles are labeled, packed and store in cool and dry places..

Flow Chart for Preparation of Grape Wine

Ripe fruits

↓

Removal of peduncle

↓

Extraction of juice by crushing or pressing

↓

Filling juice in jars up to ¾ volume

↓

Addition of sugar to make 20-25% TSS

↓

Adjustment of pH (0.6- 0..8 ACID)

↓

Addition of preservatives (Sulphur dioxide at the rate of 50–70 ppm)

↓

Keep for an hour

↓

Addition of wine yeast (20 mg/5kg grapes)

↓

Fermentation (48 hrs)

↓

Filtration

↓

Fermentation (for 10 days)

↓

Fining and filtration

↓

Aging (Maturation) for 6–8 months

↓

Pasteurization at 85°C for 2 minutes

↓

Bottling

↓

Pasteurisation at 82°C for 20 minutes

↓

Cooling

↓

Labeling and Packaging

↓

Storage.

Mixed Alcoholic–Acid Fermented Product

Vinegar

Vinegar is the mixed Alcoholic- acid Fermentation product. It is liquid obtained by Alcoholic and Acetic acid fermentation of suitable materials containing sugar and starch (at least 10% fermentable sugar).

Vinegar Production

Vinegar can be obtained from different food materials

Fruits: Grapes, Orange, Pineapple, Ripe Banana, Pear, Apricot and Jamun.

Other food materials: Vinegar can also be obtained from food materials like Potato, Barley, Honey and Coconut water.

Vinegar is produced in two steps or parts or stages of fermentation process.

Stages or Parts of Fermentation

1. First stage or step of fermentation (Alcoholic fermentation): Fruit juice is fermented with a pure culture of starter yeast. Most suitable starter yeasts are given below.
   i. Sacharomyces *ellipsoideus*
   ii. Sacharomyce *cerevisiae*
   iii. Sacharomyce *malei*

   Commonly Sacharomyces ellipsoideus is used for fermentation of fruit juice due to its rapid growth and alcohol forming power.
2. **Second stage or step of fermentation:** In the second stage, alcohol (Ethanol) is oxidized to Acetic acid and number of flavour compounds. In this stage adequate supply of air (Oxygen) is maintained.

   The changes (reactions) occur at both stages of fermentation are presented as follows.

1. Alcoholic fermentation

$$\underset{\text{(Glucose)}}{C_6H_{12}O_6} \xrightarrow[\text{Yeast}]{\text{Sacharomyces } ellipsoideus} 2C_2H_5OH+2CO_2+55\text{Kel}$$

Most suitable temperature for growth of yeast is 25-27°C.

2. Acetous fermentation

$$\underset{\text{(Ethylalcohol)}}{\text{CHOH}} \xrightarrow[\text{Bacteria}]{Acetobacter\ aaceti} \underset{\text{(Acetic acid)}}{\text{CHCOOH}}+H_2O+116\text{Kel}$$

Maturing of Vinegar

The maturing period of vinegar involves reactions between residual Ethanol (Alcohol) and Acetic acid to form Ethyl acetate which imparts characteristic flavor to the product. Vinegar contains about 5% Acetic acid has germicidal and antiseptic properties.

Strength of Vinegar

The strength of vinegar is expressed as the concentration of Acetic acid in vinegar. It is expressed as Grain.

GRAIN: Grain expresses the concentration of Acetic acid in Vinegar.

1 grain contains 0.1 g. of Acetic acid per 100 ml. vinegar at 20°C.

or

1 grain = 0.1g acetic acid/100 ml vinegar

Example: If vinegar contains 4 g acetic acid per 100ml at 20°C. means vinegar strength is 40 grains.

Types of Vinegars

There are two types of vinegar

(a) Brewed vinegar
(b) Artificial vinegar

(a) **Brewed vinegar:** It is prepared from various fruits and starchy plant materials like potato and sugar containing substances such as molasses and honey by alcoholic and subsequent acetic fermentation. Examples

i. **Fruit vinegar:** Cidar vinegar or Grape vinegar, Grape vinegar, Other Fruits used are orange, jamun, peach, pear, pineapple, and banana.

ii. **Potato vinegar**

iii. **Malt Vinegar:** From malted barley

iv. **Molasses vinegar:** It is diluted to 16% TSS neutralized with citric acid and fermentation.

v. **Honey vinegar:** Prepared from low grade honey.

(b) **Artificial vinegar:** It is prepared by synthetic acetic acid and Glacial acetic acid to a legal standard of 4% and coloured with Carmel.

Steps involved in preparation of vinegar

i. Conversion of the sugar in fruits intonversion of Alcohol by yeast called Alcoholic fermentation.

ii. Conversion of Alcohol into vinegar by Acetic fermentation (Acetification). etc.

Preparation of Vinegar

Vinegar is prepared by three processes

(a) Slow process
(b) Orleans slow process
(c) Quick process or (Generator or Grman process)

(a) Slow process: In this process fruit juice or sugar solution is filled in earthen pots or wooden barrels and kept for at least 5-6 months warm damp room to undergo spontaneous Alcoholic and Acetic fermentation. No special care is taken but mouth of containers is closed with cloth to keep out insects and dust etc.

(b) Orleans slow process: In this method clear and good quality of vinegar is prepared.

Steps involved

1. Select sweet fruits having 10% sugar in juice.
2. Peel fruits
3. Remove stone or seeds.
4. Cut into pieces and extract juice. If some fruits are hard soften fruit pieces by heating in small quantity of water.
5. Adjust sugar at least 10% in juice if sugar is low in fruit juice by adding sugar. If sugar content in juice is high then it is diluted to 10% TSS.
6. Pasteurize juice by heating to destroy undesirable microorganisms in juice.
7. Pasteurized juice is filled in containers to three fourth capacity and kept for fermentation. During fermentation all sugar is converted into alcohol which can be seen by testing with a hand refractometer which indicates 0.1% TSS. Temperature during fermentation should be 25-27°C.
8. Fermented juice is kept or stored for 2-3 weeks for sedimentation.
9. After sedimentation juice is filtered through muslin clothe.
10. Ready vinegar is kept in containers for 4-8 months for aging. During aging vinegar develops a good aroma, flavor and taste. Before aging vinegar does not have a good taste.
11. Filter or strain Caramel liquid through muslin cloth to achieve clear liquid.
12. Caramel colour is added for colouring if needed.
13. Vinegar is filled in properly washed and sterilized bottles and corked air tight.
14. Vinegar bottles are pasteurized by keeping in hot water at temperature 71-77°C. for 15-20 minutes so that further growth of bacteria is stopped.
15. Cool vinegar bottles and label and store them.

(c) Quick process

In this process additional oxygen is supplied for growth of bacteria for rapid fermentation. It is not commonly followed.

13

Preservation with Salt, Oil, Vinegar and Chutney Pickle

Preservation of food (fruits and vegetables) in common salt sand vinegar is called pickling. It is one of the most ancient method of preservation of fruits and vegetables. Pickling is the result of fermentation by Lactic acid forming bacteria which are found in large number on the surface of fruits and vegetables. They can grow in acidic medium and in the presence of 8-10 percent salt solution. The growth of majority of undesirable bacteria is inhibited. Lactic acid most active at 30 °C temperature. This temperature should be maintained during the process of pickling.

Preparation of pickles

Pickles are prepared with salt, vinegar and oil or combination of these ingredients. Spices are also used for pickle preparation.

Advantages of pickles

1. Pickles are good appetizers
2. They add palatability to meal.
3. They stimulate the flow of gastric juice and this help in digestion.

Fruits and vegetables used for pickle preparation

Fruits: Mango (unripe or immature), Lime, lemon Olive, Tind, Karaunda, Jackfruit and Aonla.

Vegetables: Carrot, Radish, Turnip, Cauliflower, Onion, Garlic and cucumber,

Types of pickles

1. Salt pickle
2. Oil pickle
3. Vinegar pickle
4. Salt, Oil and Vinegar combine pickles

1. Salt pickle

In this type of pickle salt concentration is 15% or above which prevents micro-organisms spoilage.

Role of salt

1. Salt creates unfavourable conditions for growth of microorganisms responsible for spoilage
2. Salt improves the taste and flavor of pickle
3. Salt hardens the tissues and control fermentation.

The method of preservation by salt is generally used only in vegetables in which there is little sugar content. Since sugar content is less, sufficient Lactic acid bacteria can be formed by fermentation to act as preservative. Some fruits like mango and lemon/lime are also preserved with salt.

Preparation of Mango Pickle with Salt

For preparation of mango pickle with salt the following ingredients are required. Spices in pickle are added according to taste.

Ingredients	**Quantity**
Immature or mature green fruits	1 kg
Salt	150-200 g
Spices	
Red chilli powder	10 g
Asafoetida	2 g
Fenugreek	8-10 g
Black pepper	8-10 g
Large cardamom	8-10 g
Cumin	8-10 g
Cinnamon	8-10 g
Clove	3-4 g

Procedure

1. Select raw, fresh and unripe mature healthy mango that is firm and has a clear skin.
2. Wash and dry mango thoroughly.
3. Wipe fruits with a cloth to remove water drops adhering on the skin.
4. Cut fruits in longitudinal way in to large pieces or slices with stainless steel knife.

   Make sure that cutting board, Knife, Bowl, Jar and other utensils used for making pickle are fully dry. Bowls, spoons, cutting board and knife should be stainless steel.

5. Fill the mango pieces or slices in a jar and add required quantity of common salt into slices filled in jar and mix thoroughly.
6. Keep this jar in the sun for 4-5 days or till the green colour of slice skin surface change to pale or yellow.
7. Mix all spices and roast the mixture. If needed grind the spice mixture before adding into pieces. All spices can be added without roasting also.
8. Add mixture of spices in to jar filled with mango slices and mix thoroughly.
9. Keep jars in dry and cool place for 2-3 week before use.
10. Stir mango pickle well and if needed add more salt.
11. Label and store in dry and clean place.

Flow chart for Mango Pickle with Salt

Mature green mango fruits

↓

Trimming

↓

Washing

↓

Peeling if needed

↓

Slicing

↓

sprinkling salt

↓

filling in jar

↓

Keeping in sun for one week

↓

Shaking jar to mix salt thoroughly

↓

Mixing spices

↓

Labeling

↓

Storage

(ii) Oil pickle

Oil pickle is highly acceptable and most popular particularly in India. They are highly spiced pickles. In this type of pickle fruit and vegetable slices are completely immersed in edible oil especially mustard oil. Other oils like Rye oil, Sesamum oil can also be used for pickle preparation.

Preparation of Mango Pickles with Oil

For preparation of 1 kg mango fruit with oil the following ingredients are used.

Ingredients	**Quantity**
Mango pieces or slices.	1 kg
Mustard seeds or mustard powder	30 g
Methi seeds (fenugreek seeds) coarsely grounded	125 g
Red chili powder	30 g
Asafoetida (Hing)	2 g
Common Salt (adjust as needed)	150-200 g
Garlic cloves	4 cloves
Turmeric powder.	30 g
Nigella (Kalaunji)	30 g
Black pepper	30 g
Fennel or Aniseed (Saunf)	30 g
Cumin powder	30 g
Clove	5
Cardamom powder	30 g
Mustard or Rape seed oil	350 ml or (in sufficient quantity to keep the surface covered).

Procedure

All steps mentioned above are followed in this case also.

1. Select raw, fresh and unripe mature healthy mango fruits that are firm and have clear skin.

2. Wash in clean fresh water and dry mango fruits thoroughly.
3. Wipe fruits with a cloth to make fruit skin free of water drops adhering to fruit skin.
4. Cut fruits into longitudinal way in large pieces or slices.

   Make sure that cutting board, Knife, Bowl, Jar and other utensils used for making pickle are fully dry. Bowls, spoons, cutting board and knife should be stainless steel.
5. Immerse fruit slices in 2-3 percent Brine water to prevent blackening of fruit pieces.
6. Fill the mango pieces or slices in a jar and add required quantity of common salt into slices filled in jar and mix thoroughly.
7. Keep this jar in the sun for 4-5 days or till the green colour of slice skin surface change to pale or yellow.
8. Shake jar while it is kept in the sun to mix salt properly.
9. Mix all spices and roast the mixture. If desire grind the spice mixture before adding to pieces All spices can be added without roasting also.
10. Add mixture of spices in to jar filled with mango slices and mix thoroughly.
11. Mix required quantity of mustard oil till slices are covered with oil. Oil acts as preservative.
12. Keep jars in dry and cool place for 2-3 week before use.
13. Stir mango pickle well and if needed add more salt.
14. Store this glass jar in cool place.

Care taken: Handle jar carefully while using pickle.

Flow Chart for Preparation of Mango Pickles with Oil

Mango fruits

↓

Trimming

↓

Washing

↓

Cutting into longitudinal pieces or slices

↓

Filling in jars

↓

Keeping in sunlight for a week

↓

Mixing spices after roasting

↓

Adding salt

↓

Mixing oil

↓

Labeling

↓

Storing at room temperature

Some photographs shown below are taken during class.

Selected healthy mango fruits

Cut Pieces of fruit

Mango fruit pieces kept in container.

(iii) Vinegar pickle

This preservation method is based on the addition of food grade vinegar which has a bactericidal action in concentration up to 4% acetic acid. Vegetables preserved in vinegar need to reach a final concentration of 2-3% acetic acid to assure their preservation. To achieve this 6-9% acetic acid vinegar is used as related to the specific ratio of vinegar: vegetable. If the vinegar concentration is lower than 2% vinegar in pickle need to be pasteurize to assure the preservation.

In vinegar pickle:

Salt	2-3 %
Sugar	2-5 %
Acetic acid concentration	2-3 % for 6-9 5 acetic acid vinegar is added Mango, chilli, onion, cucumber are preserved in vinegar.

Preparation of Onion Pickle with Vinegar

The following ingredients are required for 1 kg onion vinegar pickle.

Ingredients	Quantity
Onion (small size)	1 kg
Salt	250 g
Red chilli powder	10 g
Large cardamom	10 g
Black pepper	10 g
Cumin	10 g
Asafoetida	5 g
Vinegar (food grade)	1 liter.

Procedure

1. Select healthy small size onion bulbs
2. Remove outer dry skin of bulb
3. Heat in hot water at 71-77 °C for 5 minutes
4. Fill blanched onion bulbs in properly washed and cleaned jar
5. Add salt in jar
6. Keep jar for 1-2 days
7. Drain out water (exuded from onion bulbs) from jar
8. Add vinegar and spices
9. Store the jar in dry place.

Flow Chart for Preparation of Onion

Pickle with Vinegar

Onion

↓

Peeling

↓

Blanching

↓

Filling in jar

↓

Adding salt

↓

Draining water

↓

Adding vinegar and spices

↓

Store

(iv) Pickles with mixture of Salt, Oil and Vinegar

There is combine action of salt, vinegar and oil. Spices are also added for flavouring. Spices are fried in oil and mixed with prepared fruits and vegetables before addition of vinegar e.g., cauliflower pickle.

Preparation of Cauliflower Pickle

The following ingredients are required for preparation of 1 kg cauliflower curd.

Ingredients	**Quantity**
Cauliflower pieces	1 kg
Salt	150 g
Turmeric powder	15 g
Red chilli powder	15 g
Large cardamom	15 g

Garlic	10 g
Ginger pieces	25 g
Onion pieces	50 g
Rye or mustard powder	40 g
Cumin	15 g
Clove	4
Cinnamon	15 g
Tamarind	100 g
Mustard oil	150 ml
Vinegar	150 ml

Procedure

1 Select good quality of cauliflower curd (Firm, solid and creamy or milky white)
2. Trim undesirable parts of curd
3. Wash them with clean and fresh water
4. Cut into pieces in small size
5. Blanch cauliflower pieces in hot water at 71-77°C for 5 minutes
6. Dry cauliflower pieces in the sun for ½ hr
7. Fry spices in mustard oil
8. Grind spices to make powder
9. Mix spices powder with dried cauliflower pieces
10. Fill mixture in well cleaned glass jar
11. Shake the mixture in jar for thorough mixing.
12. Keep jar in sun for 3-4 days
13. Mix vinegar 60-120 ml or 3-6 ml acetic acid
14. Shake the mixture thoroughly
15. Keep jar in sun for 1-2 days
16. After that it is ready for use or storage.

Flow Chart for Preparation of Cauliflower Pickle

Cauliflower

↓

Washing

↓

Cutting into pieces

↓

Blanching in hot water for 5 minutes

↓

Drying in open sun for ½ hr

↓

Frying of spices and powdering

↓

Mixing spices in cauliflower pieces

↓

Filling in glass jar

↓

Shaking the mixture in jar

↓

Keep jar in the sun for 3-4 days

↓

Mixing vinegar 60-120 ml or 3-6 ml acetic acid

↓

Shaking mixture thoroughly

↓

Keep in sun for 1-2 days

↓

Pickle is ready for use or storage

Chutneys

Chutneys are thick mixtures made from a variety of fruits and vegetables, sugar, spices and sometimes vinegar. Any edible sour fruit can be used as a base for chutney and sugar is added to develop sweet taste. The vinegar is not added always it is added if necessary depending on the natural acidity and maturity of the fruits that is to be used. Most chutney products are boiled, which produces a syrup and alters the taste, colour and thickness. Sugar and acid present in chutney act as preservative. Spices are also added to contribute flavor and taste. They also act as preservative. Product is also pasteurized.

Characteristics of Chutney: A good quality chatney should be palatable and appetizing. Chatney may be sweet or salty or both. Sourness is desirable taste of chatney

Fruits and vegetables used for chutney preparation: Unripe mango fruit and Unripe or ripe Tomato fruits are commonly used for making different types of chatney.

Preparation of Sweet Mango Chutney

Following ingredients are needed for preparation sweet mango chutney.

Ingredients	Quantity
Mango slices or shreds	1 kg
Sugar or gur	1 kg
Salt	45 g
Onions (chopped)	50 g
Garlic (chopped)	15 g
Ginger (chopped)	15 g
Red chilli powder	10 g
Black pepper,	10 g
Cardamom (large),	10 g
Cinnamon,	10 g
Cumin,	10 g
Aniseed (powdered)	10 g
Clove (headless)	5 numbers
Vinegar	170 ml

Procedure: For preparation of mango sweet chuntey the following steps are followed.

1. Select unripe firm mango fruits
2. Wash with cold water
3. Peel with stainless steel knife or peeler
4. Grate the peeled fruits into fine slices
5. Cook slices into small quantity of water to make them soft.
6. Add sugar and salt to slices
7. Allow the mixture to stand for few hours so that sugar and salt dissolved into fruit juice exuding from the fruit slices.
8. Keep mixture on the stove or heater or Gas for cooking
9. Tie all spices in muslin cloth or bag
10. Keep bag dipped in the mixture and continue heating with continuous stirring till mixture becomes thick sufficiently
11. Remove the spices bag and squeeze into mixture
12. Add vinegar and continue heating till product gets consistency like jam.
13. Cool the product slightly and pour into clean and sterilized wide mouth bottles
14. Allow bottle to cool in the air and close bottle airtight.

Flow Sheet for Preparation of Mango Chutney

Mango fruit

↓

Washing

↓

Trimming

↓

Peeling

↓

Cutting or grating

↓

Cooking slices in little water

↓

Adding sugar and salt into boiled slices

↓

Keeping for sweetening

↓

Cooking mixture

↓

Tying all spicesin muslin cloth bag

↓

Dipping spices bag into cooking mixture

↓

Removing and squeezing spices bag

↓

Adding vinegar

↓

Continuing boiling,cooling

↓

Filling in bottles

↓

Capping bottles

↓

Storage

Other types of chutneys are also prepared in similar way.

14

Preservation by Canning

Process of preservation of food stuffs such as fruits and vegetables whole or in pieces in sugar syrup or brine by heating them in hermetically sealed containers for long storage is known as canning. In other words canning is the process of sealing food stuffs hermetically in containers and sterilizing them by heat for long storage.

Merits

1. Fruits and vegetables are canned in the season when the raw material is available in plenty.
2. Canned products are be sold in off season and given better return to growers.
3. During canning process organisms responsible for spoilage are destroyed within sealed containers by means of heat. Canning of fruits and vegetables is done under following operations or steps.
   i. **Selection:** Fruits and vegetables having uniform maturity or ripe are selected.
   ii. **Sorting and grading:** Fruits and vegetables are sorted and graded to achieve uniform quality on the basis of size and colour.
   iii. **Washing:** Fruits are generally washed thoroughly in clean running water to remove dust and spray residues. They may be washed with cold or hot water spray or by soaking. Vegetables may be soaked in solution of $KMnO_4$ to disinfect them. They are handled carefully to avoid brushing.
   iv. **Peeling:** Peeling is the removal of unwanted materials to improve appearance of the final products. Removal of skin is done by stainless steel knife to avoid darkening. It is done carefully without wasting flesh or useful parts of fruits and vegetables.
   v. **Slicing or cutting:** Fruits and vegetables are not packed in the same form as they are harvested. They are sliced or cut into pieces to reduce the size with the help of slicer or stainless steel knife.
   vi. **Blanching:** Generally fruits and vegetables are blanched by holding in hot water (85-100°C) until enzymes are inactivated. Blanching is done prior to canning. Blanching is the heat treatment given to plant

materials for inactivating and killing plant tissues to prevent enzymatic and microbial deterioration.

Why blanching is done?

1. To remove respiratory gases
2. To inhibit enzymatic reactions
3. To promote shrinkage of products
4. To hydrate certain products
5. To pre-heat products to assist vacuum formation in can.
6. **Filling product in can:** Before filling, cans are washed with hot water to remove dirt or dust. This operation is done mostly by hand.
7. **Syruping or Brining:**
8. **Syruping:** A solution of sugar in water is called syrup. Generally can is filled with hot syrup. Syruping is only practiced in fruits. Sugar syrup is measured with Refractometer.

   Preparation of syrup: There are three categories of sugar syrup depending on the requirement of fruits.

   (a) **Heavy:** Sugar is 50%

   (b) **Medium:** Sugar is 40%

   (c) **Light:** Sugar is 33%.

   Procedure for syrup preparation:

   i. Measure sugar and water according to above ration

   ii. Dissolved sugar in water

   iii. Boil the sugar solution

   iv. Filter with small cloth or muslin cloth.

   Brining: A solution of salt in water is called brine. The concentration of salt should be 26.5%, which is measured with Salometer. Brining is done only in vegetables.

   100°Salometer reading = 26.5% salt concentration.

   Objectives of syruping and brining

   i. To improve taste and flavor of canned fruits and vegetables.

   ii. To fill the inters pace between fruit and vegetable pieces.

   iii. To facilitate further heat processing or to serve as a heat transfer medium for facilitating processing. Temperature of syrup and brine at time of can filling should be 79-82°C. Head space of 0.6 cm should also be kept at the time of filling syrup or brine.
9. **Lidding or Clinching:** Placing the lid on the top of the filled can and partially seaming the lid to the can by a single roller operation action of

the double seamer is called clinching. Lid remains sufficiently loose to permit the escape of free as well as dissolved air out of the can.

10. **Exhausting the containers:** Exhausting is removal of air from the contents of containers or can.

    Objectives:

    i. To drive out air from the fruit tissues

    ii. To prevent discolouration as well as loss of flavor during storage.

    Procedure: Exhausting can be done by two methods:

    i. **Mechanical method:** It is done by machine and practiced in commercial scale production.

    ii. **Traditional method:** It is done on home scale. In this method filled cans are placed in a large kettle or open tank containing boiling water in such a way so that the top of the can is about 5 cm above the level of water in tank. Tank is covered by a lid and heating of water continues till temperature in center of the can reach 80-82°C. It will take about 10 minutes.

11. **Sealing:** In case if exhausting is done mechanically cans are sealed air tight immediately after exhausting by sealing sealer.

    In case if exhausting is done traditionally in boiling water, cans are removed one by one while still hot from hot water and lids are air tightly sealed by a home can sealer. During sealing temperature should not fall below 70°C.

12. **Pasteurization:** In case of home scale, cans are placed on a thick pad of cloth spread at the bottom of the pan containing hot water and continue heating till water comes to rolling boil. This step is necessary for killing microorganisms present in the content of the can

13. **Cooling:** After exhausting process, cans are cooled by dipping in tank containing cold water. Cans are also cooled quickly in running cold water and taken out when they are still warm. Cans are dried in air before storage.

14. **Storage:** Cans are stored in dry and cool place.

Flow Sheet for Canning

Fruit/Vegetables

Sorting

↓

Grading

↓

Washing

↓

Peeling

↓

Slicing

↓

Blanching

↓

Filling

↓

Syruping or Brining

↓

Lading or Clinching

↓

Exhausting

↓

Sealing

↓

Pasteurizing

↓

Cooling

↓

Storage

15

Value Addition and Products Developed by Fruits and Vegetables

Several products of fruits and vegetables are prepared on commercial and home scale. Products development from fruits and vegetables is the best way to utilize excess produces and save them from post harvest losses..

Value Addition

This is the process of improving the value of a commodity through processing by changing its form, increasing its shelf life or better presentation like packaging. Value addition can be done through the ways.

1. Changing the form of the fruits and vegetables,
2. Improving the presentation of the them or their products,
3. Increasing their shelf-life.

Importance of Value Addition

1. Increase income through the products which are of a higher value than the original forms.
2. Improve storability
3. Differentiate taste of food
4. There is easy transportation because value added products are not bulky which reduces the cost of transportation.
5. Improve presentation through changing packaging

Products Development of Fruits and Vegetables

Important products developed from fruits and vegetables are given below:

1. **Fruit juices:** Juice is a pure fruit product which may be natural or sweetened.
2. **Squashes and cordials**

   Squash: This is a type of fruit beverage and diluted before serving

   Cordials: It is simply crystal-clear squashes.
3. **R. T. S:** It is a diluted form of fruit beverage which is consumed without dilution.

4. **Syrups:** Syrups are filtered juices that are concentrated by boiling until the sugar content reaches 50-70%. The heat and high solids content preserves the syrup and it is used in place of sugar or honey. Syrups can be made from a wide range of fruits, but the most common type is made from grapes
5. **Jam:** Jam is a solid gel made from fruit pulp or juice from a single fruit or from a combination of fruits. The sugar content is normally 68-72%, which will prevent mould growth.
6. **Jelly:** Jellies are crystal-clear jams that are produced using filtered juice instead of fruit pulp containing more fruit pectin.
7. **Marmalade:** Marmalades are produced mainly from clear citrus juices (including lime, orange, grapefruit, lemon and orange) and have fine shreds of peel suspended in the gel. Ginger may also be used alone or in combination with the citrus fruits. The fruit content should not be less than 20% citrus fruit and the sugar content is similar to jam. The correct combination of acid, sugar and pectin is needed to achieve the required gel structure and rapid boiling is necessary to remove water quickly to concentrate the mixture before it darkens and loses its ability to form a gel.
8. **Fruit Cheese:** Fruit cheeses are fruit pulps that are boiled until they have a final sugar content of 75-85%. They set as a solid block without the need for added pectin, but they do not have the same gel structure as jams and marmalades. They can be cut into bars or cubes to eat directly.
9. **Fruit preserves:** It is a product prepared by keeping a whole mature fruit or its large pieces in sugar syrup till it becomes tender, transparent and soluble solid concentration is achieved about 68% or more.
10. **Candied fruits/vegetables:** It is a product prepared by impregnation of sugar syrup into fruit and vegetable and subsequently drained free of syrup followed by surface drying to give a transparent coating of sugar on the surface. The total sugar content of the impregnated fruit is kept at about 75% to prevent fermentation. The left syrup can be used for the preparation of next batch of product or in Pickles, Sauce, Vinegar making.
11. **Fruit pickles:** Preservation of fruits and vegetables in common salt sand vinegar is called pickling. It is one of the most ancient method of preservation of fruits and vegetables. Pickling is the result of fermentation by Lactic acid forming bacteria which are found in large number on the surface of fruits and vegetables.

    Important Hints of Preparation of Pickles

    a. Pickles can be prepared without use of oil by adding lemon juice particularly in chilli, carrot, cauliflower, pickles while by adding vinegar in mango, jack-fruit, chilli, cauliflower.

b. Vinegar acts as a preservative in pickles. In order to ensure good result the final percentage of acid should be below 2.

c. Salt acts a preservative in vegetable pickles. Spoilage is prevented adding 15-20 % of common salt. 4. Lime pickle should consists of 12 % of common salt salt and 1.2 % acidity

12. **Fruit Vinegar:** Vinegar is the mixed Alcoholic- acid fermentation product. It is liquid obtained by Alcoholic and Acetic acid fermentation of suitable materials containing sugar and starch (at least 10% fermentable sugar).

13. **Grape Wine:** It is fermented juice of grape containing alcohol up to 11-14% but it may be as low as 7%. alcohol. Wine is produced by alcoholic fermentation of Grape or Grape juice by yeast (*Saccharomyces serevisiae*).

Different value added products of fruits and vegetables are prepared by processing. Common and important home scale products are described below. The quantity of ingredients given may vary according to taste and flavour. The ingredients of some products given are based on author experience and knowledge gain from home and working in the Post harvest technology of fruits and vegetables project.

A. Fruit Products

Mango Products

Different products of both green as well as ripe mango fruits can be preprepared.

1. Green Mango Fruits

Utilization of green fruits dropped down: It is a common problem of flower and fruit drop in mango. Generally when there is high wind or rain with high wind this problem occurs. There is big question in some countries how to utilize these dropped green mature fruits properly specially in Ethiopia and Eritrea. This is my observation during my stay in these countries. In the country like India most of dropped green fruits are utilized properly for preparation of the following products.

A. Preparation of Dried Mango Pieces or Slices

Dried mango slices are used for developing taste and flavor of Dal especially Arhar (Peagon pea).

Procedure

Steps are taken for preparation of dried mango slices.

1. Collect dropped green fruits.

2. Select uniform size fruits
3. Wash them with fresh clean water.
4. Trimming: Stem attached with fruits and injured damaged parts of the fruits are removed.
5. Peel the fruits i. e. removal of skin of fruits.
6. Cut fruits in to large pieces with stainless steel knife in longitudinal way.
7. Coring or removal of seed. At this stage in most of fruits stone is not formed if formed only seed is removed.
8. Dip cut mango fruit pieces in salt water for few minutes to avoid browning of the cut fruit pieces.
9. Remove cut fruit pie4ces from salty water.
10. Spread these fruit pieces on cloth on the well cleaned floor in the open sun light for drying.
11. Cover these fruit pieces by a thin cloth to avoid dust particles contamination.
12. These fruit pieces will become dry within 4-5 days
13. Dried fruit pieces are now ready for packing and they are packed in cleaned dried wide mouth bottles or containers or cloth bags.
14. Store packed dried mango fruit pieces in dry and cool places at room temperature.

B. Preparation Ofdried Mango Powder (Amchur)

Dried mango powder is used for good adjustment in ice cream, bakery and confectionary industry. For preparation of Amchur all above mentioned steps are taken. One extra step is taken in the preparation of Amchur that these dried mango pieces are crushed or grinded in grinding machine to make powder. This powder is packed in bottles and labeled.

Flow Chart for Mango Dried Slices

Green mature mango fruits

↓

Washing

↓

Peeling

↓

Cutting fruit slices or piece

↓

Dipping into 2% salt solution (2g/lt. water)

↓

Draining out salt solution

↓

Dipping slices in 2000 ppm SO2 solution for 2 hrs.

↓

Spread slices in thin layers in wooden trays or cloth

↓

Dry slices in Sunlight (fresh and dried slices ratio is 6:1).

↓

Slices can be dried in Solar dryer

↓

Fill slices in well cleaned dry air tight container

↓

Labeling

↓

Store in dry place

C. Preparation of Mango Pickles

Pickles are prepared with salt, oil vinegar, mixture of salt, oil, vinegar and spices. There are several kinds of pickles sold in the market. Mango pickle ranks first. Fruit pickles are generally preserved in sweetened and spices vinegar while vegetable pickles are preserved in salt. In India, oil pickles are highly popular

Preparation of Mango Pickles with oil

Steps for preparation of mango pickles have been described already with pickle preparation. Dropped green fruits of mango can be utilized for pickle preparation.

The following ingredients are needed for mango pickle preparation with oil.

Ingredients	**Quantity**
Mango pieces	1 kg,
Salt	150 g,
Fenugreek (powdered)	25 g,
Turmeric(powdered)	15 g,
Nigella seeds	15 g,
Red chilli powder	10 g,
Clove (headless)	8 numbers,
Black pepper,	15 g
Cumin,	15 g
cardamom (large),	15 g
Aniseed (powdered) each	15 g,
Asafoetida	2 g,
Mustard oil	350 ml (sufficient to cover pieces)

Procedure: Different steps have been given earlier with pickle preparation.

Flow Chart for Preparation of Mango Pickles with Oil

Mango fruits

↓

Labeling

↓

Trimming

↓

Washing

↓

Cutting into longitudinal pieces or slices

↓

Filling in jars

↓

Keeping in sunlight for a week

↓

Mixing spices after roasting

↓

Adding salt

↓

Mixing little oil if needed

↓

Labeling

↓

Store at room temperature

Preparation Mango Pickle with Salt

Ingredients	**Quantity**
Mango slices	1 kg
Salt	200 g
Red chilli powder	10 g
Cnnamon	10 g
Asafoetida	5 g
Cumin	10 g
Cardamum	10 g
Black pepper	10 g
Cloves	6
Fenugreek or Methi	10 g

Procedure: As described in pickle preparation earlier. Both types of pickles are used with food or meal at any time.

Preparation of Mango Sweet Chutney

Mango chatney is prepared by green mature fruits. It is used with meal, snacks and breads. It's constituents are salt, fruit, chiilli, some green spice like coriander and sugar.

Ingredients	Quantity
Mango slices	1 kg
Salt	45 g
Sugar	1 kg
Onion (chopped)	50 g
Red chilli	10 g
Cumin	10 g
Garlic (chopped)	15 g
Cinnamon	10 g
Cardamom (large)	10 g
Fennel (Aniseed) powder	10 g
Ginger (chopped)	15 g
Cloves	6
Black pepper	10 g
Vinegar (good quality)	170 ml

Procedure: Steps have already been described earlier.

Flow Chart for Preparation of Green Mango Fruit Sweet Chutney

Selection of unripe fruit

↓

Washing

↓

Peeling

↓

Grating into slices

↓

Cooking slices

↓

Adding sugar and salt to slices

↓

Allow them to sweeten and dissolved sugar and salt

↓

Heating mixture

↓

Dip spices tied in muslin cloth or bag in the mixture

↓

Continue heating and stirring

↓

Remove spices bag and squeeze into mixture

↓

Add vinegar

↓

Continue heating ttencyill the product get jam consistency

↓

Cool product slightly

↓

Pour product into sterilized wide mouth bottles

↓

Cool it in air and close air tight

↓

Label

↓

Storage

Preparation of Mango Fruit Preserve and Candies (Murabba)

Ingredients	Quantity
Fruit slices or pieces	1 kg
Sugar	1.5 kg
Water	little water

Procedure: Following steps taken are taken.

1. Select uniform mature fruits
2. Wash them with fresh and clean water.
3. Keep them in water for overnight or 12 hrs.
4. Remove them from water and keep for drying of surface water in the shade or room temperature.
5. Peel them to remove the skin
6. Cut peeled mango fruits into large pieces in cube form.
7. Prick them by stainless steel fork.
8. Take some fresh clean water in a stainless steel bowl and add some salt in water.
9. Dip fruit pieces in the salty water for overnight. Take into mind that the amount of water should be sufficient so that fruit pieces can be dipped wholly.
10. Remove fruit pieces from salt water next morning.
11. Wash mango fruit pieces at least two times with fresh clean water.
12. Blanch the fruit pieces in boiling in water for 3-4 minutes.
13. Remove them from hot water.
14. Put blanched fruit pieces in stainless steel bowl
15. Take a wide stainless steel bowl and equal quantity of sugar to the weight of fruits (1 kg fruit + 1 kg sugar.)
16. Put a layer of sugar first then second layer of fruit pieces like wise put sugar and fruit pieces in layers one by one. Keep for a day
17. During this period, fruit juice from fruit pieces will ooze out and dissolve sugar. Sugar will enter into fruit pieces. The whole sugar will become watery in solution form.
18. Add citric acid in required quantity or lime juice to the mixture of fruit pieces and sugar.
19. Keep on regulated electric heater or Gas for boiling. Let it boil for 10-15 minutes.
20. No need to add preservative because mixture contains more than 70 percent sugar which will not allow microbes to survive in the mixture.

21. When sugar syrup become thick consistency check the end point whether it is ready or not.
22. For this take a drop of syrup in a tea spoon and cool it
23. Take this drop between the finger and thumb and separate gently both. If it is ready a thread like structure will be formed, if not continue boiling till it is observed.
24. After observing end point close the gas and remove bowl from the Gas.
25. Keep it for cooling, now it is ready.
26. Fill the ready preserve (Muraba) in well cleaned wide mouth glass bottles.
27. Store bottles at cool and dry place for future use.

Step for Candy Preparation

1. After testing syrup thick consistency
2. Boiling is stopped and boiling materials is cooled
3. Drain out sugar syrup
4. Spread in single layer of boiled fruit pieces on flat plats for drying
5. Keep stand pieces for 48 hrs till a transparent shining layer of sugar developed on the fruit surface
6. When candy is dried fill in clean jars
7. Close jar
8. Label it
9. Store in dry and cool place.

Flow Sheet for Preparation of Mango Preserve/Candy

Mature fruit

↓

Washing

↓

Trimming

↓

Peeling

↓

Cutting into slices or pieces

↓

Blanching

↓

Cooling

↓

Forking

↓

Putting forked pieces and sugar in alternate layers

↓

Keeping stand for 48 hrs

↓

D raining sugar syrup

↓

Adding citric acid or lime juice and little water into sugar syrup

↓

Boiling mixture

↓

Putting fruit pieces into boiling syrup

↓

Heating continuously till considerable thick consistency is observed

↓

Testing end point

↓

Draining out syrup

↓

Spreading fruit pieces on flat plat

↓

Keeping stand for 48 hrs to become dry

↓

Filling dried candies in clean sterilized jars

↓

Close jar

↓

Labeling

↓

Storage

Preparation of Green Mango Squash

Ingredients:

Materials	**Quantity**
Green mango slices	1 kg
Sugar	1.6 kg
Water	1.5 liter
Salt	40 g
Mint leaves (if required)	250 g
Cumin	10 g
Black pepper powder	4 g
Citric acid or lime juice	5 g
Sodium benzoate	1g/kg pulp

Procedure

The following Steps are involved in preparation of green mango fruit squash.

1. Select healthy fruits
2. Wash fruits with fresh and clean water
3. Peel fruits to remove skin
4. Remove stone
5. Cut fruits into pieces
6. Add water equal to weight of mango fruits

7. Heat mango pieces to soften them
8. Crush and extract pulp or juice
9. Filter mango juice
10. Prepare sugar syrup mixture with spices and salt
    a. Mix required quantity of sugar, salt, lime juice or citric acid, cumin and black pepper and other spices needed.
    b. Boil the mixture
    c. Filter the sugar syrup mixture through muslin cloth.
11. Mix mango juice with sugar syrup mixture
12. Heat the mixture
13. Add preservative (Sodium benzoate) in required quantity
14. Fill ready squash in cleaned sterilized bottles leaving 1.5-2.0 cm space at head
15. Close bottles with cork or cap
16. Label bottles and store in dry and cool place.

Flow Sheet for Preparation of Green Mango Squash

Green mango fruits

↓

Washing

↓

Trimming

↓

Peeling

↓

Cutting into slices or pieces

↓

Blanching

↓

Crushing to make pulp

↓

Strain juice or pulp

↓

Mix sugar, salt, citric acid and all spices

↓

Boil mixture

↓

Filter sugar syrup mixture

↓

Mix mango juice with sugar syrup mixture

↓

Add preservative

↓

Heat mixture

↓

Fill into clean sterilized bottles

↓

Capping bottles

↓

Cooling

↓

Labeling

↓

Storage

2. Ripe Mango Fruits

Many products are developed from ripe mango fruits.

I. Preparation of Mango Pulp

Ingredient:

Materials	**Quantity**
Ripe Fruit juice	1 kg
Sugar Sufficient quantity (according to taste)	
Potassium Meta bisulphate (Preservative)	610 mg/kg pulp

Procedure

1. Select fresh uniform fully ripe fruits and discard over ripe and juicy fruits and rotten fruits.
2. Wash fruits thoroughly with fresh and clean water.
3. Trim the stem portion attached with fruit
4. Soften the fruits by pressing them with palm
5. Squeeze out the pulp in stainless steel bowl or pan
6. Filter the pulp through a course muslin cloth to remove pieces of fibers.
7. Add sufficient quantity of sugar according to taste to sweeten the pulp.
8. Mix thoroughly the mixture
9. Add preservative in the mixture
10. Heat the mixture
11. Fill hot pulp in cleaned dry and sterilized bottles.
12. Seal bottles and
13. Keep bottles in boiling water for 20 minutes
14. Cool them in running water
15. Label the bottle
16. Store bottles in dry place at room temperature.

Flow Sheet for Preparation of Mango Pulp

Mango fruits

↓

Selection of fruits

↓

Washing in clean and fresh water

↓

Trimming

↓

Pressing fruits with palm to make soft

↓

Squeeze out pulp

↓

Filtering of pulp

↓

Adding of sugar in pulp according to taste

↓

Mix sugar with pulp thoroughly

↓

Filling in bottles

↓

Sealing or capping of bottles

↓

Keeping bottles in boiling water for 20 minutes

↓

Cooling bottles in running water

↓

Labeling of bottles

↓

Storage

(II) Preparation of Mango Juice

Ingredients

Materials	Quantity
Fruit pulp	1 kg
Sugar	200 g
Citric acid.	1 g
Water	800 ml

Procedure

The following steps are involved in preparation of this product.

1. Select fresh uniform fully ripe fruits and discard over ripe and juicy fruits and rotten fruits.
2. Wash fruits thoroughly with fresh and clean water.
3. Trim the stem portion attached with fruit
4. Peel the fruits with stainless steel knife
5. Cut and separate the pulp and stone
6. Keep pulp pieces in stainless steel bowl or pan
7. Crush pulp
8. Filter or strain the pulp through a course muslin cloth to remove pieces of fibers
9. Preparation of sugar syrup
    a. Add water at the rate of 0.8 liter/kg pulp and required quantity of sugar and citric acid
    b. Mix mixture thoroughly and heat.
10. Mix sugar syrup with pulp and preservative potassium meta bisulphite at the rate 610mg/kg pulp.
11. Heat the mixture at 85°C for 15-20 minutes.
12. Fill into bottles or can
13. Seal the bottles by lid.
14. Keep in boiling water 100°C for 20 minutes.
15. Cool them at room temperature.
16. Label the bottles
17. Store them.

Flow Sheet for Preparation of Mango Juice

Mango fruits

↓

Selection of fruits

↓

Washing in clean and fresh water

↓

Trimming

↓

Peeling

↓

Cutting into pieces and separation of stone and pulppieces

↓

Keeping pulp pieces into stainless steel bowl

↓

Crushing of pulp pieces

↓

Filtering pulp through course muslin cloth

↓

Preparing sugar syrup

↓

Mixing sugar syrup + pulp + preservative

↓

Heating mixture for 15-20 minute at 85°C

↓

Filling into bottles or cans

↓

Sealing of bottles

↓

Keeping bottles in boiling (100 °C) water for 20 minutes

↓

Cooling bottles at room temperature

↓

Labeling of bottles

↓

Storage

(III) Preparation of Mango Squash

Ingredients

Materials	**Quantity**
Fruit pulp	1 kg
Sugar	1.7 kg
Water	1 liter
Citric acid	20 g
Potassium Metabisulphite.	610 mg/kg pulp

Procedure

Steps taken for preparation of mango squash have been described earlier. Flow chart for preparation of mango squash is given earlier.

(IV) Preparation of Mango RTS

Ingredients

Material	**Quantity**
Fruit pulp	1 kg
Sugar	600g
Citric acid	10 g
Edible yellow colour if available	0.2 g
Water	3.5 lters or kg

The following steps are taken to prepare R. T. S (Ready To Serve) beverages of Mango.

Procedure has already been given earlier. However, steps described for preparation of this beverage.

Steps

1. Take fully ripe fruit and avoid juicy fruits.
2. Wash fruits with fresh clean water
3. Remove the stem part attached with mango fruit.
4. Soften the fruit by pressing with hand palm ao that pulp can be extracted easily.
5. Extract the pulp in a stainless steel bowl..
6. Weigh the fruit pulp according to required quantity (1 kg of fruit pulp)
7. Take required quantity of sugar, citric acid and water (600 g+ 10 citric acid + 3.4 liter water) in a stainless steel bowl.
8. Mix them thoroughly till they are dissolved.
9. Add fruit pulp and edible colour to the sugar mixture.
10. Mix them properly.
11. Fill mixture in 200-300 ml well cleaned and sterilized bottles leaving 2-3 cm neck space.
12. Beverage is ready to serve immediately.
13. If it is kept for future consumption, then dip bottles in water filled in a vessel.
14. Heat slowly by raising temperature up to 75°C. for 20-30 minutes.
15. Remove bottles from hot water
16. Cool them at room temperature.
17. R. T. S. is ready to keep for future consumption up to 3-4 months period.

Flow chart for preparation of R. T. S. is given before previous chapter.

(V) Preparation of Mango Jam

Ingredients

Material	**Quantity**
Fruit pulp	1 kg
Sugar	750 g
Citric acid (Lime juice)	5 ml
Preservative KMS (Sodium benzoate or Potassium metabisulphite).	2 g/kg pulp
Water	250 ml

Procedure

The following steps are taken for preparation of mango jam.

1. Healthy and uniformly ripen fruits were selected
2. Selected fruits were washed properly with fresh and cleaned water.
3. These fruits were peeled with the help of stainless steel knife.
4. Fruit flesh was removed from stone with the help of stainless steel knife.
5. Required quantity of sugar and citric acid, preservative (sodium benzoate) were added with fruit flesh
6. Mixture was put on regulated electric heater for boiling.
7. Boiling mixture was stirred continuously for proper boiling.
8. It was boiled till a thick consistency observed in juice mixture.
9. End point of jam was determined by sheet test method.
10. When it is ready it was taken down from the heater.
11. Hot jam was filled in the cleaned sterilized bottles.
12. Lid was kept on mouth of the bottle.
13. Filled bottles were cooled at room temperature and lid was tied
14. Bottles were labeled with a containing all detail of jam.
15. These jam bottles were place in dry and cool place for storage.

Flow Sheet for Preparation of Mango Jam

Ripe fruits

↓

Washing

↓

Trimming

↓

Peeling

↓

Cutting into pieces or slices

↓

Crushing or squeezing to make pulp or juice

↓

Straining pulp through muslin cloth

↓

Pouring in stainless steel vessel

↓

Mixing sugar, Lime juice and preservative into pulp

↓

Boiling mixture till thick consistency observed

↓

Testing end point

↓

Filling ready jam in clean sterilized bottles

↓

Capping or corking bottles

↓

Cooling

↓

Labeling

↓

Storage

(VI) Preparation of Mango Leather (Amawat)

Ingredients;

Materials	**Quantity**
Mango pulp	1 kg
Potassium metabisulphite	2 g per kg pulp (In villages mango Amawat is prepared without addition of preservative)

Procedure

Similar initial steps taken for extraction of mango pulp are followed.

1. Select fresh uniform fully ripe fruits and discard over ripe and juicy fruits and rotten fruits.
2. Wash fruits thoroughly with fresh and clean water.
3. Trim the stem portion attached with fruit
4. Soften the fruits by pressing them with palm
5. Squeeze out the pulp in stainless steel bowl or pan
6. Filter the pulp through a course muslin cloth to remove pieces of fibers.
7. Take the required quantity of mango pulp in a stainless steel bowl
8. Add required quantity of preservative
9. Mix it with pulp thoroughly.
10. Take stainless steel tray or plate (Traditionally clean cloth sheet is used)
11. Put butter or Ghee in tray and smear it to avoid sticking of pulp with tray surface.
12. Spread pulp in thin layer
13. Cover tray with net cloth or muslin cloth to avoid contamination of dust.
14. Put tray in sun for drying
15. When first layer is dry put the second layer of pulp upon it. Repeat the process till 0.6-1.25 cm thick layer is not formed.
16. Let it dry
17. Cut mango pulp layer into small pieces with the help of stainless steel knife.
18. Wrap them into butter paper and pack them into bottles.

Flow Chart for Preparation of Mango Leather (Amawat)

Mango fruit

↓

Selection of fruit

↓

Washing in clean and fresh water

↓

Trimming

↓

Pressing fruits with palm to make soft

↓

Squeeze out pulp

↓

Filtering of pulp

↓

Taking pulp in bowl

↓

Add preservative and mix thoroughly

↓

Take stainless steel plate

↓

Putting and smearing of butter on the inner surface

↓

Spreading pulp in thin layer on the plate surface

↓

Keep plate in sun for drying

↓

When first layer dried put second layer of pulp

↓

Repeat the process till 0.6-1.25 cm thick layer is not formed.

↓

Cover plate or tray with a thin cloth to avoid dust

↓

Let it dry

↓

Cutting of pulp layer in to small pieces

↓

Wrapping dried pieces in butter paper

↓

Packing in bottle

↓

Store in dry place

(VII) Preparation of Mango Shakes for Immediate Use

It is prepared by fruit pulp mixing with milk and sugar.

Ingredient: The quantity of ingredients can be taken according to the taste.

Materials	**Quantity**
Milk	1 kg
Ripe mango fruit pulp	500 g
Sugar	100 g (according to taste)
Cardamom	3-4
Water	200 ml
Ice cubes	3-4 per drinking glass (according to requirement)

Procedure: Following steps are taken for preparation of mango shake which is consumed immediately.

1. Take 1 liter (kg.) boiled and cooled milk in a stainless steel vessel
2. Take well ripe firm fruits,
3. Peel them
4. Cut pulp slices with stainless steel knife to separate the stone
5. Put mango slices into Mixture and Grinder
6. Crush mango slice into pulp
7. Take mango pulp in a stainless steel vessel
8. Pour milk in pulp and mix thoroughly
9. Add sugar and cardamom powder into milk and pulp mixture.
10. Mix this mixture thoroughly
11. Put ice cubes into mixture and stir it so that drink becomes cold. It is

ready to use. If child mango shake is needed put it in the refrigerator for few minutes.

12. Pour mango shake into drinking glasses for use.

Flow Sheet for Preparation of Mango Shake

Ripe fruits

↓

Trimming

↓

Washing

↓

Peeling

↓

Separating pulp from stone

↓

Mixing pulp and milk thoroughly with mixer

↓

Adding sugar and cardamom powder into mixture

↓

Mixing thoroughly

↓

Adding ice cubes to make cold

↓

Serving

Citrus Products

Orange

1. Preparation of Orange Squash

Ingredients

Materials	**Quantity**
Orange juice	1 liters
Sugar	500 g
Water	500 ml
Lemon juice	2-3 tsp
Potassium Meta bisulphate	½ tsp

Procedure

The following steps are involved for preparation of orange squash.

1. Select healthy orange fruits
2. Wash fruits with clean and fresh water
3. Peel out skin
4. Extract juice with juice extractor
5. Take juice in a stainless steel
6. Sugar syrup preparation
   (a) Mix sugar, citric acid and water
   (b) Heat till sugar and citric acid are dissolved
   (c) Cool sugar syrup
   (d) Filter syrup through a muslin cloth.
7. Blend or mix clean syrup with fruit juice
8. Add preservative either Sodium benzoate or Potassium Metabisulphite (KMS) in a small quantity of boiled and cooled water.
9. Heat or boil the mixture
10. Cool the squash
11. Add permitted colour and essence in adequate amount if required
12. Fill the squash in dry sterilized bottles leaving 2.5 cm head space.
13. Close bottles with cork or cap
14. Label bottles
15. Store in dry and cool place

Flow Sheet for Preparation of Orange Squash

Orange fruits

↓

Washing

↓

Trimming

↓

Peeling

↓

Extraction of juice

↓

Strain juice

↓

Prepare sugar syrup (Mix sugar, salt, citric acid and water)

↓

Boil mixture

↓

Filter sugar syrup mixture

↓

Mix orange juice with sugar syrup mixture

↓

Add preservative

↓

Heat mixture

↓

Add edible colour and essence if required

↓

Fill into clean sterilized bottles

↓

Capping bottles

↓

Cooling

↓

Labeling

↓

Storage

Lime/Lemon

I. Preparation of Lime/Lemon Squash

Ingredients

Materials	**Quantity**
Lime/lemon juice	1.0 kg
Sugar	1.7 kg (Quantity of sugar is more because of ore sourness)
Citric acid rich in citric acid	(Not required because of more acidic)
Potassium Meta bisulphate	¾ tsp (350 ppm)
Water	1 kg or liter

Procedure

The steps followed during the teaching are given below.

1. Healthy ripe fruits are selected
2. Selected fruits are washed with clean and fresh water.
3. Fruits are peeled with help of stainless steel knife.
4. Juice was extracted by Juicer.
5. Juice was measured with help of measuring cylinder.
6. Sugar syrup was prepared by adding sugar I Water in required proportion as given above. No need to add citric acid because Lime/Lemon contain sufficient quantity of citric acid.

7. Mixture of sugar and water was added in juise
8. This mixture was boiled by putting on heater.
9. Boiling was stirred and scams was removed.
10. Add preservative with juice and mixed properly.
11. Squash was kept for cooling
12. Ready hot squash was filled in well cleaned and sterilized bottles
13. Bottles are kept at room temperature for cooling.
14. Labeling of bottles
15. Cooled bottles are stored in dry and cool place.

Flow chart has already been given earlier.

II. Preparation of Lime and Green Chilli Pickle

Ingredients

Materials	Quantity
Fruit	1 kg
Green chili fruits	1 kg
Salt	250 g

Procedure

Following steps are involved in preparation of this product.

1. Take healthy and ripe lime fruits
2. Wash them with fresh and clean water
3. Cut fruits into halves or quarters pieces longitudinally
4. Keep fruit pieces into a sterilized well cleaned glass bottles or jar
5. Take healthy green chilli fruits
6. Wash them with clean and fresh water
7. Trim fruit stalk attaché with fruit
8. Keep in a vessel or bowl
9. Mix lime fruit pieces and chilli fruits together
10. Squeeze some lime fruits and mix juice into fruit mixture
11. Mix salt and sugar (if required) with fruit pieces and chilli mixture. The quantity of sugar should be according to taste.
12. Mix some spices to develop taste
13. Mix this mixture thoroughly
14. Fill mixture of pickle in a jar
15. Keep jar for some days so that the skin of fruits will become soft.

16. When skin is soft then it is ready to use.
17. Label it
18. Store in dry place

Flow Chart for Preparation of Lime and Chilli Pickle

Lime and Chilli

↓

Selection of lime and chilli fruits

↓

Washing in clean water

↓

Trimming

↓

Cutting of lime fruits into pieces

↓

Mixing both lime pieces and chilli fruits together in bowl

↓

Mixing salt + sugar (if required for taste) +some spices (for taste and flavour)

↓

Mixing salt + sugar + spices with fruit mixture

↓

Filling this mixture in glass jar

↓

Squeezing some lime juice in the mixture

↓

Mixing Thoroughly by shaking

↓

Liding or capping (sometimes a cloth piece is tied on the mouth of jar)

↓

Keeping jar filled with pickle for few days

↓

Skin of fruits become soft

↓

Labeling

↓

Storage

III. Preparation of Lime Pickle With Salt

Ingredients

Materials	**Quantity**
Lime	1 kg,
Salt	200 g,
Red chilli powder	15 g,
Cinnamon,	10 g
Cumin,	10 g
Cardamom (large)	10 g
Black pepper (powdered)	10 g
Clove (headless)	5 numbers

Procedure

Steps take for preparation for lime pickle with salt

1. Take healthy and ripe fruits
2. Wash them with fresh and clean water
3. Cut fruits into halves or quarters pieces longitudinally
4. Keep fruit pieces in to a sterilized well cleaned glass bottles or jar
5. Squeeze some lime fruits and mix juice into fruit pieces
6. Mixt salt and sugar (if required for taste) with fruit pieces.
7. Mix some spices and red chilli powder to it

8. Mix thoroughly this mixture.
9. Keep jar for some days so that the skin of fruit will become soft.
10. When skin is soft then it is ready to use.

Flow Chart for Preparation of Lime/Lemon Pickle In Salt

Fully ripe fruits

↓

Trimming

↓

Washing

↓

Cutting into pieces

↓

Squeezing of juice from ½ amount of fruits

↓

Mixing spices and salt with juice

↓

Mixing lime pieces

↓

Filling in clean jar

↓

Covering with lid

↓

Keep in sun for 4-6 days

↓

Store in dry and cool place

IV. Preparation of Lemon and Lime Pickle with Oil

Ingredients

Materials	**Quantity**
Lime fruits	1 kg
Salt	120 g
Red chilli powder	10 g
Turmeric powder	10 g
Cumin	10 g
Cardamom	10 g
Black pepper powder	10 g
Asafoetida	2 g
Mustard oil	500 ml

Procedure

Pickle in oil is also prepared as procedure given above for lime pickle with salt.

Flow Chart for Preparation of Lime/Lemon Pickle in Salt

Fully ripe fruits

↓

Trimming

↓

Washing

↓

Cutting into pieces

↓

Squeezing of juice from ½ amount of fruits

↓

Mixing spices and salt with juice

↓

Mixing lime fruit pieces

↓

Filling in clean jar

↓

Add oil with mixture

↓

Covering with lid

↓

Keep in sun for 4-6 days

↓

Store in dry and cool place

Banana Products

A. Green Banana Fruits

I. Preparation of Banana Chips

Ingredients

Required quantities of different ingredients are taken according to need.

i. Unripe or green mature fruits
ii. Cooking oil
iii. Salt
iv. Black pepper powder

Procedure

Steps taken for Banana chips preparation are as follows:

1. Select healthy green banana fruits
2. Wash banana fruits with fresh and clean water
3. Peel fruits
4. Cut fruits into slices (not more nor very thin) with stainless steel knife
5. Keep slices in 5% salt + 0.25% citric acid water for 10 minutes to avoid browning
6. Blanch slices in hot water for 5 minutes
7. Keep slices in 1% KMS solution for 30 minutes.

8. Keep them in trays in the open sun light or Solar drier for 48 hrs for drying. Sometime slices are spread in a single layer on clean cloth for open sun drying
9. Cover the slices with a thin muslin cloth to protect from dust contamination. Slices are protected when they are dried in Solar drier.
10. When slices become dried they are packed in containers like glass jars.
11. Capping of glass jar
12. Label the container filled with slices
13. Store in dry place
14. Fry slices before use
15. Spread black pepper powder on fried slices to develop taste
16. Consume chips

Flow Chart for Preparation of Banana Chips

Selection of fruits

↓

Washing

↓

Peeling

↓

Slicing

↓

Keeping in salty water for 10 minutes to avoid browning or blackening

↓

Blanching in hot water for 5 minutes

↓

Dipping slices in 1% KMS solution for 30 minutes

↓

Spreading slices on cloth or in trays

↓

Drying of slices in sunlight or Solar dryer

↓

Packaging dried slices in containers (glass jar)

↓

Capping

↓

Labeling

↓

Storage

↓

Frying chips when used

↓

Spreading black pepper powder

↓

Consumption

(B) Ripe Banana Fruits

II. Preparation of Banana Milk Shake

Ingredients: Quantity of ingredients are take according to the amount of Banana shake needed

i. Ripe fruits

ii. Milk

iii. Sugar

iv. Water

Procedure: Steps taken for preparation of Banana milk shake are described as follows:

1. Select healthy ripe banana fruits
2. Wash them with clean water
3. Peel them to remove skin
4. Cut into slices
5. Take sugar according to sweetness needed

6. Mix sugar with banana slices
7. Mix milk in this mixture
8. Take this mixture into mixture and grinder to grind and mix thoroughly
9. Add edible rosewater or small cardamom powder to develop flavor and taste
10. Add ice cubes before consumption to cool it. If chilled banana shake is needed keep in refrigerator for few minutes
11. Pour in drinking glasses
12. Drink is ready for use.

Flow chart for preparation of banana milk shake is the similar to preparation of Mango milk shake.

Papaya Products

(A) Ripe Fruits

I. Preparation of Papaya Jam

Ingredients

Materials	**Quantity**
Papaya fruit pulp	1 kg
Sugar	750 g
Citric acid	5 g
Water	250 ml
Preservative KMS or Sodium Benzoate	2g/kg fruit pulp

Procedure: The following steps are taken for preparation of Papaya jam.

1. Healthy and uniformly ripe fruits are selected
2. Selected fruits are washed properly with fresh and cleaned water.
3. These fruits are peeled with the help of stainless steel knife.
4. Peeled fruits are cut into slices or pieces with help of stainless steel knife
5. Fruit slices or pieces are crushed to make pulp with the help of ladle or stainless steel knife or spoon.
6. Required quantity of sugar, water, citric acid and preservative are added with fruit pulp and mix thoroughly
7. This mixture is put on regulated electric heater for boiling.
8. Boiling mixture is stirred continuously for proper boiling.
9. Boiling is continued till a thick consistency observed in pulp mixture.

10. End point of jam is determined by sheet test method.
11. When it is ready it is taken down from the heater.
12. Hot jam is filled in the cleaned sterilized bottles.
13. Lid is kept on mouth of the bottle.
14. Filled bottles are cooled at room temperature and lid is tied
15. Bottles are labeled with a containing all detail of jam.
16. These jam bottles are placed in dry and cool place for storage.

Flow Chart for Preparation of Papaya Fruit Jam

Fully ripe fruits

↓

Washing with clean water

↓

Peeling

↓

Cut into slices or pieces

↓

Pulpping

↓

Mix sugar, citric acid to fruit pulp

↓

Heating or boiling till thick consistency received

↓

Testing end point

↓

Filling in bottles

↓

Cooling

↓

Capping or corking

↓

Labeling

↓

Storage

II. Preparation of Papaya Fruit Candy

Papaya candy is also prepared as pumpkin candy is prepared.

Procedure: Steps taken are as follows.

1. Select a healthy mature and ripe papaya fruit
2. Washed fruits with fresh clean water
3. Peel fruit with a sharp stainless steel knife carefully
4. Peeled fruit was divided into pieces an
5. Seeds are removed and inner fruit surface is cleaned properly to remove seed attachments
6. Cut fruit into large cube like pieces
7. Cubes of fruit were forked with stainless steel fork or wooden fork or bamboo pricker
8. Take sugar equal quantity of fruit weight
9. Take a stainless steel bowl having flat bottom
10. Spread a first layer of sugar in bowl and put one layer of fruit cubes. Similarly put sugar and cubes in alternate layers
11. Let it stand for over night
12. Fruit juice will ooze out and sugar will be dissolve in juice and penetrate in fruit cubes
13. Drain out sugar syrup
14. Sugar syrup is boiled
15. 5 ml lime juice per kg weight of fruit are added in boiling sugar syrup and mixed thoroughly.
16. Scam produced in sugar syrup is removed.
17. Fruit cubes are put in syrup gently because syrup is hot and continue boiling till sugar syrup become bit thick
18. For testing end point, the boiling syrup is taken by dipping a stainless steel spoon and cool it. A drop of syrup was taken between finger and thumb and finger and thumb were separated gently. If a thread like

structure of the syrup occurs then it is ready and if not the boiling of syrup is continued till thread like structure is not received.

19. When it is ready it was removed from the heater and stopped boiling and kept for cooling.
20. Take out cooked fruit cubes from the syrup with the help of fork or spoon and keep on wide plates for drying.
21. Cubes are kept for drying for about 48 hrs or more and covered with muslin cloth to avoid dust.
22. Change the sides of cubes with help of spoon or fork which for proper drying.
23. When fruit cubes become dry a shining layer of syrup is formed on the surface of fruit cube. Then it is dried and ready
24. Ready candies are filled in properly washed and well dried bottles or jars
25. Lid is put and tied properly
26. Bottles are labeled mentioning name of fruit candy and other details if needed.
27. Bottles are stored in dry and cool place.

Flow chart for papaya fruit candy is similar to Pumpkin fruit candy

Grapes Products

Preparation of Grape Juice

Ingredients

Materials	**Quantity**
Ripe grape fruits	1 kg
Water	50 ml
Sugar	50 g per liter jjuice
Sodium benzoate	1 g per kg juice

Procedure: Steps taken for preparation of grape juice are mentioned below.

1. Select good quality of fruits
2. Remove stem and discard under ripe and decayed fruits
3. Wash fruits with clean water to remove dust adhering to fruit surface
4. Take about 50 ml water in a pan and put grape in it
5. Heat them till fruits become soft so that fruits can be crushed easily to extract juice and cannot slip away when pressure is applied.
6. Press gently and extract juice
7. Measure the juice

8. Add sugar at the rate of 50 g per liter juice to adjust desirable taste depending on the sweetness of the varieties of grape
9. Fill the juice mixture in a jar or container and keep it in refrigerator for 8-10 days. This will facilitate thick mass to settle down.
10. Decant the juice and filter juice through muslin cloth.
11. Heat juice to 90 °C.
12. Add preservative
13. Fill hot juice in cleaned and sterilized bottles
14. Pasteurize bottles at 85 °C. for 15, 20 and 30 minutes according the size of bottles.
15. Cool bottles at room temperature
16. Label bottles
17. Store bottles in dry and cool place.

Flow chart for preparation of grape juice is similar to the preparation of other fruit juice.

Jackfruit Products

A. Ripe Jackfruit

I. Preparation of Jackfruit Squash

Ingredients

Materials	Quantity
Fruit pulp	1 kg
Sugar	1.7 kg
Water	1.3 kg
Citric acid	40 g
Potassium metabisulphite	600 mg/kg pulp.

Procedure: Following steps are taken.

1. Select fully ripe and healthy fruits
2. Cut fruit into eight or sixteen pieces first across and then lengthwise
3. Remove gummy core by a sharp knife to free the bulbs
4. Smear little cooking oil on the hand to prevent sticking.
5. Separate the bulbs from the rind and surrounding carpels
6. Trim top and bottom of the bulb to remove seeds and their covering.

7. Cut the bulbs into halves, quarter or slices.
8. Weigh them and crush the bulb pieces
9. Take half weight of water of weight of bulb pieces and heat slowly to soften them.
10. Mash soften pieces of bulbs to make pulp by crushing them thoroughly.
11. Rub the pulp through a fine sieve to remove tiny large pieces.
12. Prepare sugar syrup by mixing sugar in water and dissolving by heating at 60 °C.
13. Add citric acid and preservative K MS at the rate of 600 mg per kg of pulp.
14. Remove scam and filter syrup
15. Mix well sugar syrup with pulp and heat mixture
16. Squash is ready.
17. Fill it in well cleaned, sterilized and dried bottles.
18. Cool bottles at room temperature.
19. Label them and
20. Store in dry and cool place.

Flow Chart for Preparation of Jack fruit Squash

Ripe fruit

↓

Smearing hand with oil

↓

Cutting fruit into pieces

↓

Removing gummy core to make free bulbs

↓

Separating bulbs from rind and carpels

↓

Trimming upper and lower part of bulbs

↓

Cutting bulbs into slices or pieces.

↓

Weighing slices

↓

Softening of slices

↓

Crushing and mashing soften slices to make pulp

↓

Rubbing pulp through a fine sieve to remove tiny large pieces of pulp

↓

Preparation of sugar syrup

↓

Addition of citric acid and preservative KMS

↓

Mixing sugar syrup and pulp

↓

Heating mixture

↓

Filling in bottles

↓

Labeling

↓

Storage

B. Unripe Jackfruit

II. Preparation of Jackfruit Pickle In Oil

Ingredients

Materials	**Quantity**
Fruit Slice	1 kg
Salt	65 g
Red chilli powder	20 g
Ginger (chopped or crushed)	2 g
Onion (chopped)	30 g
Mustard seeds	3-4 g
Cumin	3-4 g
Clove	3-4
Cardamom	2 g
Cinnamon	3-4 g
Garlic (crushed)	3 g
Vinegar	125 ml
Sugar	125 g

Procedure: Following steps are taken for preparation of Jackfruit pickle with oil.

1. Select green immature fruit in which seeds and pithy portion have not become hard.
2. Wash fruit thoroughly with clean and fresh water
3. Smear oil in hand to prevent sticking of jack fruit milky liquid
4. Trim the stem portion
5. Peel the thick outer rind with a stainless knife
6. Cut the peeled fruit into large pieces (1.5 - 2.0 cm thick)
7. Fill them into clean jar and add 8 percent salt solution
8. Keep slices or pieces immerse in brine
9. Following day add more salt to raise brine to 10 percent
10. Keep process continue to 8-10 day
11. Repeat it daily to raise salt concentration up to 15 percent.
12. During this process pieces become soft and clean
13. Cured slices are washed twice with clean and fresh water
14. Fry chopped ginger, onion, garlic and mustard seeds in little oil until they become brown.

15. Add this rest of spices to the fruit pieces and put in jar.
16. Mix well and keep jar in sun for 3-4 days
17. Stir occasionally contents of jar
18. Cover mouth with cloth piece to prevent house flies.
19. Add sugar and vinegar in the jar
20. Mix well and keep again in the sun for 3-4 days.
21. After that pickle is ready for use
22. Fill it in jar and mix sufficient oil to cover the pickle.
23. This helps to improve quality of pickle.
24. Store in dry place

III. Preparation of Jackfruit Pickle with Vinegar

Ingredients

Materials	**Quantity**
Salt cured fruit slices	1 kg
Sugar	150 g
Vinegar	200 ml
Water	75 ml
Spices	
Clove	3-4g
Cardamom	3-4 g
Coriander	3-4g
Ginger	2 g
Mustard seeds	30 g

Procedure: Steps followed for preparation of Jackfruit pickle with Vinegar are given below.

1. Tie all spices in a muslin cloth
2. Take vinegar, sugar and water in a vessel
3. Place spices in vessel containing vinegar, sugar and water
4. Cook gently on low fire for 3-4 hrs at 50°C to extract spices in mixture
5. If needed more spices may be used according to taste
6. Allow vinegar extract to cool slowly and remove cloth bag
7. Mix salt cured fruit slices and spiced vinegar
8. Vinegar pickle is ready.
9. Label jar
10. Store in dry place

Flow chart for preparation of pickle has been already been given previously.

Pineapple Products

Preparation of Pineapple Squash

Ingredients

Materials	**Quantity**
Juice	1 kg
Sugar	1.7 kg
Citric acid	42 g
Water	1.3 kg
Potassium metabisulphite	2.4 g

Pineapple esscence and colour are added if needed.

Procedure: Steps followed for preparation of pineapple squash are mentioned below.

1. Select ripe fruit
2. Wash fruits with clean and fresh water
3. Trim crown
4. Peel fruit with a sharp stainless steel knife
5. Remove eyes and damaged portion of fruit with stainless steel knife
6. Cut fruit into small pieces
7. Chop them finally with a sharp stainless steel knife.
8. Wrap chopped fruit in muslin cloth
9. Press the cloth containing fruit pieces to extract juice
10. Strain juice with coarse muslin cloth
11. Mix ingredients juice, sugar, water and citric acid and stir thoroughly
12. Strain this mixture through a thick muslin cloth
13. Heat Mixture at 85-90°C. to dissolve ingredients sugar and preservative.
14. Fill it in squash bottle which is well cleaned and sterilized.
15. Cool it at room temperature.
16. Store bottle in dry and cool place.
17. When use take one cup of squash and add 3 cups of water

Flow chart for Pineapple squash is similar to preparation of other fruit squashes

Sapota Products

I. Preparation of Sapota Squash

Ingredients

Materials	**Quantity**
Fruit pulp	1 kg
Sugar	1.2 kg
Water	1.0 kg
Citric acid	40 g
Potassium metabisulphite	600 mg/kg pulp
or Sodium benzoate	710 mg

Procedure

Following steps are taken.

1. Select ripe soft fruits
2. Wash fruits with clean and fresh waters
3. Peel fruits with stainless steel knife
4. Cut fruits into segments and remove seeds
5. Pass the fruit segments through a meat mincer to obtained uniform pulp,
6. Wrap minced segments in a thick muslin cloth and press out the juice with small basket press.
7. Mix sugar, water and citric acid and heat mixture to make syrup
8. Strain syrup through a muslin cloth
9. Mix filtered syrup and pulp
10. Mix preservative to the product
11. Heat the mixture at 85 °C.
12. Fill product into cleaned sterilized bottles
13. Close the mouth of bottles by tiding cap
14. Cool them and
15. Store bottles in dry and cool place

Flow chart is similar to other fruit squash

II. Preparation of Sapota Jam

Ingredients

Materials	**Quantity**
Fruit pulp	1 kg
Sugar	1.5 kg
Citric acid	15 g
Water	250 g
Sodium benzoate	710 mg/kg product
or Potassium metabisulphite	600 mg per kg product

Procedure

Steps taken are mentioned below:

1. Select ripe and soft fruits
2. Wash fruits with clean water
3. Peel fruits with stainless steel knife
4. Cut fruits into segments and remove seeds
5. Pass the fruit segments through a meat mincer to obtained uniform pulp,
6. Wrap minced segments in a thick muslin cloth and press out the juice with small basket press.
7. Mix sugar, water and citric acid and heat mixture to make syrup
8. Strain sugar syrup through a muslin cloth
9. Mix filtered sugar syrup and pulp
10. Mix preservative to the product
11. Cook the mixture as given in mango jam preparation
12. Test end point by method given with jam preparation whether jam is ready or not
13. Fill product into cleaned sterilized wide mouth bottles
14. Close the mouth of bottles by tiding cap
15. Cool them at room temperature
16. Store Jam bottles in dry and cool place

Flow chart is similar to mango jam preparation

Aonla Products

I. Preparation of Aonla Candy

Ingredients

Material	**Quantity**
Aonla Fruits	1 kg
Sugar	1 kg
Water	300 ml
Citric acid or Lime juice	5 ml

Procedure

Following steps are taken for preparation of Aonla candy:

1. Select ripe and healthy fruit
2. Wash fruit with fresh and clean water
3. Fruits are forked with stainless steel fork or wooden fork
4. Dip forked fruits in 2 percent salt solution for 24 hrs
5. Wash fruits with clean water again
6. Dip fruits in 2 percent Alum solution for 24 hrs
7. Wash fruits again with clean water
8. Blanch fruits in boiling water for 5 minutes to soften them
9. Take a flat bottom stainless steel bowl
10. Put first layer of sugar and second layer of fruit and put sugar and fruit layers alternately.
11. Keep it for overnight. Fruit juice will ooze out from fruits and sugar will be dissolved. Sugar will penetrate into fruits.
12. Drain out sugar syrup and take into another bowl and mix 250 ml of water and 5 ml of lime juice.
13. Boil sugar mixture on heater and after boiling few minutes, scam formation occurs.
14. Remove scam produced on the sugar syrup surface
15. Stir syrup sometime for proper boiling
16. Put fruit in sugar syrup gently because syrup is hot
17. Boil sugar syrup with fruit till it become bit thick
18. Test end point of boiling syrup by dipping a stainless steel spoon and cool it. A drop of cooled syrup is taken between finger and thumb and finger and thumb are separated gently. If a thread like structure of the syrup occurs then it is ready and if not it has to be boiled till thread like structure is not received

19. When it is ready, it was removed from the heater and stopped boiling and kept for cooling.
20. Take out cooked fruit from syrup with the help of fork or spoon and keep on wide plates in layer for drying for about 48 hrs and covered with muslin cloth to avoid dust.
21. Chang the sides of fruits with help of spoon or fork for proper drying.
22. When fruit become dry, a shining layer of syrup is formed on the surface of fruit. Then it is dried and ready
23. Ready candy is filled in properly washed and well dried bottles or jars
24. Lid is put and tied properly
25. Bottles are labeled mentioning name of fruit candy and other details if needed.
26. Bottles or jars are stored in dry and cool place.

Flow chart for Aonla candy is similar to preparation Pumpkin candy

II. Preparation of Aonla Pickle

Ingredients

Materials	**Quantity**
Fruit	1 kg
Salt	150 g
Turmeric powder	10 g
Red chilli powder	10 g
Nigella seeds	10 g
Fenugreek powder	10 g
Clove (headless)	8 numbers,
Black pepper,	10 g
Cumin,	10 g
Cardamom (large),	10 g
Aniseed (powdered)	10 g,
Asafoetida	2 g
Mustard oil	350 ml

Procedure

Following steps taken for preparation of Aonla pickle are given below.

1. Select healthy and fully mature fruits
2. Wash them with clean water
3. Remove stem portion attached with fruits and damaged and infected parts
4. Soften fruits in hot water at 85°C. for 10 minutes
5. Separate fruit segments and seeds

6. Dry in the sun for 2 hrs to remove moisture on the surface of fruit pieces.
7. Mix all spices and fry in oil.
8. Mix thoroughly fried spices with fruit segments. If needed make powder of spces
9. Fill this mixture in well cleaned and dry jar
10. Pour oil sufficiently to cover fruit segments
11. Pickle is ready
12. Label the jar
13. Store in dry and cool place.

Flow Chart for Preparation of Aonla Pickle

Select fully mature fruit

↓

Washing

↓

Trimming

↓

Blanching in hot water for 10 minutes

↓

Separation of fruit segments and seeds

↓

Keep in the sun to dry moisture for 2hrs

↓

Mix ingredients and mustard oil with dried fruit segments

↓

Fill in well cleaned and dry jar

↓

Label jar

↓

Store it for future use

Karaunda Products

I. Preparation of Karaunda Pickles

Ingredients

Materials	**Quantity**
Mature unripe Fruits	1 kg
Salt	200 g
Red chilli powder	10 g
Turmeric powder	10 g
Cinnamon	10 g
Fennel	10 g
Cumin	10 g
Cardamom (large)	10 g
Black pepper	10 g
Cloves	6
Fenugreek or Methi	10 g
Mustard oil	200 ml

Procedure

As described in pickle preparation earlier.

Flow Sheet for Preparation of Karonda Pickles with Oil

Mature unripe fruits

↓

Washing

↓

Trimming

↓

Cutting into two half pieces and remove seeds

↓

Mixing salt with fruit pieces

↓

Filling in jars or bottles

↓

Keeping in sunlight for 4 days

↓

Mixing spices

↓

Fry spices in oil

↓

Mix fried spices with fruit pieces thoroughly

↓

Labeling

↓

Storing at room temperature

II. Preparation of Kauronda Fruit Candy

Karonda candy is also prepared as given with other fruits

Procedure

Following steps are taken for preparation of Kauronda candy.

1. Select mature unripe fruits
2. Wash fruits with fresh and clean water
3. Fork fruits with stainless steel fork or wooden fork
4. Take a stainless steel bowl with flat bottom
5. Put first layer of sugar in bowl
6. Put second layer of forked fruits. Whole sugar and fruits are put in alternate layers.
7. Keep the bowl for overnight. Fruit juice will ooze out and dissolve sugar. Sugar will penetrate into fruits.
8. Next day drain sugar syrup and take in separate pot for sugar syrup preparation.
9. Mix sugar syrup + 250 ml of water +1 g citric acid or 5 ml lime juice.
10. Keep sugar syrup mixture on heater for boiling few minutes so that syrup mix properly and scams produced are removed.
11. Put forked fruit in sugar syrup gently because syrup is hot
12. Continue boiling mixture till sugar sugar syrup becomes bit thick
13. Test end point by taking boiling syrup. Dip a stainless steel spoon and

cool it. A drop of syrup is taken between finger and thumb and finger and thumb are separated gently. If a thread like structure of the syrup occurs then it is ready and if not then it has to be boiled till thread like structure is not received.

14. When it is ready it is removed from the heater and stopped boiling and keep for cooling.
15. Take out cooked fruit from the syrup with the help of fork or spoon and keep fruit on wide plates in a single layer for drying.
16. It is kept to dry for about 48 hrs and covered with muslin cloth to avoid dust.
17. Change the sides of fruit pieces with help of spoon or fork for proper drying.
18. When fruit pieces become dry a shining layer of syrup is formed on the surface of fruit.
19. Now candy is ready. Ready candies are filled in properly washed and well dried bottles or jars
20. Lid is put and tied properly
21. Bottles are labeled mentioning name of fruit candy and other details if needed.
22. Bottles are stored in dry and cool place.

Flow chart is similar with preparation of other fruit candy.

Phalsa Products

Preparation of Phalsa Juice

Ingredients

Materials	**Quantity**
Ripe fruits	1 kg
Water	0.5 kg
Sugar	if needed for sweetening
K. M. S	600 mg per liter juice

Procedure

Similar steps like other fruit squash are followed.

Flow Sheet for Preparation of Phalsa Juice

Selection of fully ripe fruits

↓

Washing

↓

Trimming

↓

Addition of water

↓

Crushing

↓

Heating on up to 85 °C. with continues crushing

↓

Straining to separate the seeds and skin

↓

Juice

↓

Pasteurization

↓

Filling in cans or bottles

↓

Capping and Sealing

↓

Store in dry place

Ber Products

I. Preparation of Ber Fruit Candy

Ber fruit candy is prepared as the procedure given for Pumpkin candy. Different steps are shown in flow sheet given below.

Flow Sheet of Preparation of Ber Candy

Selection of fruits

↓

Washing

↓

Trimming

↓

Pricking

↓

Disinfecting

↓

Hot water treatment for 5 minutes for softening

↓

Sulphuring (2g/kg) for 2hrs.

↓

Preparation of sugar syrup
(75 percent of fruit weight +1 percent citric acid + 250 ml water)

↓

Dipping pricked fruits in sugar syrup for 24 hrs

↓

Cooking

↓

Draining out syrup from fruits

↓

Drying of fruits on wide plates for 48 hrs

↓

Packaging

↓

Storage

II. Preparation of Dried Ber Fruits

Generally wild Ber fruits are dried and available in market in many countries like India and Eritrea, Dried fruits are eaten as such by people. Especially children like it more. Drying methods for fruits and vegetable has already been described. Drying steps are shown in flow sheet below.

Flow Sheet for Drying of Ber Fruits

Selection of fruits

↓

Washing

↓

Trimming

↓

Pricking

↓

Blanching in hot water for 5 minutes

↓

Sulphuring at the rate of 2g/kg of sulphur for 2 hrs

↓

Drying

↓

Packaging

↓

Storage

III. Preparation of Dried Ber Powder

Dried fruit powder is prepared from dried fruits similarly the procedure given with dreid mango powder and given in flow sheet below.

Flow Sheet for Preparation of Dried Ber Powder

Selection of Ber fruits
↓
Washing
↓
Trimming
↓
Removal of seeds
↓
Cutting into small pieces
↓
Blanching in hot water for 5 minutes.
↓
Sulphuring (2g/kg of sulphur) for 2 hrs.
↓
Drying
↓
Grinding to make Powder
↓
Sieving through 60 mesh
↓
Packaging
↓
Labeling
↓
Storage

Bael Products

I. Preparation of Bael Fruit Pulp

Bael pulp is also prepared in similar way like mango pulp. However, following process is given for its preparation in flow sheet below.

Flow Sheet for Preparation of Fruit Pulp

Select ripe fruits

↓

Break the shell

↓

Separate pulp, seeds and fibers

↓

Add little water with pulp

↓

Crush pulp by hand and mix thoroughly with water

↓

Filter through muslin cloth

↓

Add preservative (Sodium benzoate)

↓

Pasteurize pulp

↓

Fill in well cleaned sterilized bottles

↓

Capping

Labeling

↓

Storage

II. Preparation of Bael Sharabat

Sharabat is also prepared as pulp is prepared. Only more quantity of water and sugar is added. It is a diluted form of pulp which is used immediately specially during summer season, because it has cooling effect.

Ingredients

Materials	**Quantity**
Fruit pulp	1 kg
Sugar	According to taste because Bael pulp is very sweet
Water	1 liter
Lime juice	5 ml
Sodium benzoate	0.5 g/kg

Procedure

Steps taken are given below.

1. Select well ripe fruits
2. Wash with fresh water
3. Break hard skin of fruit with the help of heavy stone piece or stainless steel pastel
4. Separate pulp from the hard skin with stainless steel knife
5. Keep in stainless steel bowl or vessel
6. Add water into pulp
7. Smash pulp with hand to separate seeds from pulp and mix pulp with water
8. Strain through muslin cloth
9. Mix lime juice into pulp
10. Add sugar according to taste and mix thoroughly
11. It is ready for immediate use but if it is kept for future use then add preservative
12. Pasteurize mixture at 85 °C. for 5 minutes
13. Fill in clean and sterilized bottles or jars
14. Close bottles with cap
15. Cool bottles
16. Label bottles
17. Store them in dry and cool place

Flow Sheet for Preparation of Bael Sharabat

Ripe fruits

↓

Washing

↓

Breaking hard skin

↓

Peeling

↓

Cutting into pieces

↓

Crushing

↓

Mixing water

↓

Smashing pulp and mixing with water thoroughly

↓

Staining pulp

↓

Adding lime juice, sugar and preservative into pulp

↓

Mixing thoroughly

↓

Pasteurizing

↓

Filling into sterilized bottles

↓

Cooling

↓

Labeling

↓

Storage

III. Preparation of Bael Preserve or Muraba

Ingredients:

Materials	**Quantity**
Fruit pulp pieces	1 kg
Sugar	1 kg
Lime juice	5-10 ml

Procedure

Different steps are shown in flow sheet below.

Flow Sheet for Preparation of Bael Preserve

Select ripe fruit

↓

Break hard shell

↓

Cut into eight pieces

↓

Blanch them

↓

Prick fruit pieces with stainless steel fork

↓

Prepare sugar syrup

↓

Mix fruit slices with sugar syrup

↓

Boil this mixture for 5 minutes

↓

Allow to stand overnight

↓

Add more sugar to raise Brix up to 70 percent

↓

Boil mixture again till Bix reaches 72 percent.

↓

Fill final product in well cleaned sterilized wide bottles or jars

↓

Label jars

↓

Store them in dry place

Mulberry Products

Ripe fruits are used for preparation of fruit squash

Preparation of Mulberry Fruit Squash

Ingredients

Materials	**Quantity**
Ripe fruits	1 kg
Sugar	500 g
Lime juice	5 ml
Sodium benzoate	0.5 g
or	
Potassium metabisulphite	

Procedure

Following steps are taken for preparation of Mulberry squash.

1. Select uniformly ripe fruits

2. Wash fruits with clean and fresh water
3. Take washed fruits in a muslin cloth
4. Crush fruits with stainless steel spoon or knife
5. Press crushed fruits with hand to extract fruit juice
6. Extract juice in a stainless steel vessel
7. Weigh fruit juice
8. Add sugar (55 percent of fruit juice weight) + Lime juice)
9. Mix ingredients thoroughly
10. Add preservative Sodium benzoate or KMS.
11. Pasteurize it by heating at 85°C. for 5 minutes
12. Fill squash in cleaned and stelized bottles.
13. Close bottles with crown cork
14. Cool bottles at room temperature
15. Label bottles
16. Store at cool and dry place.

Flow Sheet for Preparation of Mulberry Fruit Squash

Ripe fruits

↓

Washing

↓

Trimming

↓

Crushing

↓

Extracting juice

↓

Measuring juice

↓

Mixing juice, lime juice and KMS

↓

Pasteurizing at 85 °C for 5 minutes

↓

Filling in bottles

↓

Corking

↓

Cooling bottles

↓

Labeling

↓

Storage

B. Vegetable Products

A number of products are prepared from different vegetables.

1. **Sauce and Ketchup** - Tomato and Chilli
2. **Pastes and purees** - Tomato, Garlic, Ginger.

   Pastes and purees can be made from any fruit or vegetable. The most common vegetables are tomato, Ginger and garlic, which are widely used in cooking. These products can be made at a small scale by evaporating water carefully to concentrate the pulp. Constant stirring is done to prevent darkening or localized burning. The concentration of solids in the paste is normally around 36%. The high solids content and natural acidity are sufficient to preserve the product for several days but pasteurization in bottles or cans is needed for a longer shelf life. In some preparations, sugar, salt and vinegar are added to assist in preservation
3. **Pickles**

   (a) **Oil and spices pickles** - Chilli, Carrot, Radish and Turnip

   (b) **Vinegar, Salt and spices pickle** - Cauliflower,

   (d) **Vinegar pickle** - Onion
4. **Dried vegetables:** Besides these products, vegetables are also dried and preserve for future use. Cauliflower, Red ripe chilli, Fenugreek, Chana Sag.

Tomato Products

Number of products can be processed from tomato fruits. For products development following points are to be considered:

1. Uniformly ripened, red fruits are selected,
2. All green blemished and over-ripe fruits are discarded.
3. A good quality juice should be of deep red colour and possess the characteristic taste and flavour of tomato, containing about 0.4 per cent acid (in terms of citric acid)
4. Juice should be uniform in appearance and, have high nutritive value
5. Tomato juice/pulp can be extracted by hot or cold pulping. Hot pulping is superior to cold pulping because in the latter case, extraction of juice is somewhat difficult and its yield is less, vitamin C is oxidized more rapidly, the juice is lighter in colour and there are chances of microbial spoilage.
6. Extracted tomato juice in homes is strained through a steel sieve or muslin cloth

1. Tomato Juice

Preparation: Tomato juice is extracted either by cutting into pieces or fruits are soften in hot water and smash to make pulp. In the home it is prepared by using following ingredients.

Ingredients

Materials	**Quantity**
Juice	1 litre
Sugar	100 g
Salt	5 g
Citric acid and	1 g
Sodium benzoate	1 g

(If juice is used in future the add preservative)

Procedure

Following steps are taken.

1. Select fully uniformly red ripe fruits
2. Wash fruits with clean and fresh water
3. Remove peduncle and diseased fruits are discarded
4. Extraction of juice–juice is extracted either by heating and softening skin.

In first method

(a) Fruits are heated for few minutes in boiling water to soften them

(b) Drain out water and cool fruits

(c) Peel skin with a sharp stainless steel knife

(d) Cut into pieces

(e) Smash fruits with hand or stainless steel spoon

(f) Strain pulp through muslin cloth to remove seed.

In Second method

(a) Cut into small pieces

(b) Put fruit pieces in a stainless steel vessel

(c) Crush pieces with wooden ladle

(d) Cook the crushed mass at a temperature 80-85 °C. till the skin loosened from flesh.

(e) Strain mass through muslin cloth or 1 mm mesh stainless steel sieve by rubbing gently.

5. Mix juice, salt and sugar
6. Boil the juice mixture
7. Pour hot mixture in cleaned and sterilized (boiling water) bottles
8. Seal bottles air tight with cork sterilized in boiling water for 2-3 minutes.
9. Pasteurize the bottles by placing them in a vessel and cover with water
10. Boil water in which bottles are immerged in vessel for ½ hr.
11. Remove bottles from boiling water in vessel
12. Cool bottles
13. Label bottles
14. Store them in dry and cool place

2. Tomato Sauce/Ketchup

There is no much difference between sauce and ketchup.

Tomato sauce: Tomato sauce is prepared by strained pulp or juice i. e. without seed and fruit skin and it has generally thin consistency and more total solids than tomato ketchup about 30%.

Tomato ketchup: Tomato ketchup preparation includes pulp and seeds etc. It has thick consistency than tomato sauce and less total solids about 28%.

Tomato Sauces and tomato ketchups are prepared from the similar ingredients such as tomato juice or pulp and spices, salt, sugar and vinegar, with or without onion and garlic in the same manner.

Preparation of tomato sauce and tomato ketchup

Ingredients

Materials	Quantity
Tomato juice or pulp	1 kg
Sugar	75 g
Salt	10 g
Onion (chopped)	10 g
Garlic (chopped)	5 g
Clove	5
Cardamom	10 g
Cinnamon	10 g
Black pepper	10 g
Cumin	10 g
Red chilli powder	5 g
Vinegar	25 ml
Glacial acetic acid/Lime juice	5 ml (if needed)
Sodium benzoate	0.25 g

Procedure

Tomato sauce and Tomato ketchup are prepared by the same method or manner. Following steps are taken for preparation of both tomato sauce and tomato ketchup.

1. Select fully and uniformly ripe red coloured fruits
2. Wash fruits with fresh and clean water
3. Remove peduncle and discard diseased fruits
4. Cut into small pieces with stainless steel knife
5. Put fruit pieces in a stainless steel vessel or bowl
6. Crush pieces with wooden ladle or spoon
7. Cook the crushed mass at a temperature 80-85°C for 10 minute till the skin loosened from flesh. Sometime, fruits are cooked without cutting into pieces to loosen the fruit skin and when fruits become cool then skin is separated from pulp by hand and fruits crushed to make fruit mass. For preparation of tomato ketchup this procedure is followed.
8. Strain mass through muslin cloth or 1 mm mesh stainless steel sieve by rubbing gently
9. Discard the seeds and skins in the case of Sauce
10. Add sugar and preservative into pulp
11. All required spices are tied in a cloth bag.
12. Immerse spices bag in the pulp

13. Cook the mixture till the original volume of pulp becomes half.
14. Remove the spices bag and squeeze bag into pulp so that spices extract is added in the pulp
15. Add required quantity of vinegar and salt into the pulp
16. Boil the mass for few minutes on regulated electric heater or gas so that final product can be achieved. The original volume of mass becomes half and now product is ready.
17. Fill the ready product into clean sterilized sauce bottles and leave 1.5 2.0 cm neck space.
18. Cool bottles at room temperature
19. Label bottles with all informations.
20. Store bottles in dry and cool place

Flow Sheet for Preparation of Tomato Sauce and Ketchup

Ripe fruits

↓

Washing

↓

Extraction of pulp or juice

↓

Boiling juice and adding sugar and Preservative to juice

↓

Continuing heating or boiling

↓

Immersing or putting spice bag in juice

↓

Removing and squeezing spices bag

↓

Adding Vinegar and Salt

↓

Filling hot sauce in clean and sterilized bottles

↓

Corking or capping

↓

Cooling

↓

Labeling

↓

Storage

3. Tomato Chatney

Tomato Chatney is prepared in the same way as tomato sauce is prepared. It may be sweet or salty. Generally tomato sweet Chatney is preferred and prepared from ripe tomato fruits, whereas, salty Chatney is mostly prepared by mature green or turning ripe fruits.

Preparation of Tomato Sweet Chatney

Ingredients

Materials	**Quantity**
Tomato fruits	1 kg
Sugar	500 g
Salt	25 g
Onion chopped	100 g
Ginger	10 g
Garlic	5 g
Red chilli powder	10 g
Cinnamon	10 g
Cardamum	10 g
Black pepper powder	10 g
Cumin	10 g
Sodium benzoate	0.5 g
Vinegar (Glacial acetic acid)	100 ml

Procedure

Steps taken for preparation sweet tomato chatney are given below.

1. Select healthy and fully ripe uniformly red fruits
2. Wash them thoroughly with clean fresh water
3. Dip them in boiling water for one minute to loosen the fruit skin
4. Remove fruits from boiling water quickly and cool in cold water. This will facilitate to remove skin easily
5. Peel fruits by hand and keep in stainless steel bowl or vessel
5. Crush peeled fruits with help of wooden ladle or spoon
7. Spices such as cinnamon, cardamom, black pepper and cumin are taken in a pan and roasted. They are grinded before added to juice
8. Mix juice or pulp, salt, sugar, coarsely grinded spices and other spices.
9. Boil mixture to obtained thick consistency
10. Add vinegar and cook for 10 minutes again
11. Fill hot chatney in wide mouth bottle
12. Seal the bottle and tight cork
13. Label jar
14. Store in cool and dry place

4. Tomato Paste and Puree

Tomato Puree: Tomato Puree is a concentrated pulp containing 9-12% tomato solids and prepared with or without adding salt. On the basis of concentration of pulp tomato Puree can be grouped into two groups medium tomato puree containing not less than 9.0% tomato solids and heavy tomato puree containing not less than 12% tomato solids.

Tomato Paste: If tomato pulp is concentrated more and containing not less than 25 per cent tomato solids is known as tomato paste. If it is further concentrated to 33 per cent or more of solids, it is called concentrated tomato paste.

Tomato pulp is prepared from ripe tomatoes in the same manner as tomato juice.

Tomato juice with pulp is cooked in pan or open cooker for concentration so that juice can have 14-15% tomato solids. While cooking a little butter or edible oil is added to avoid sticking, burning and foaming. The end point of paste and puree is determined either by hand refractometer or by measuring the volume of tomato juice and volume of final product with the help of a measuring stick. If the concentration of final product is higher than required then add more juice. If it has low concentration then boil more to achieve the required concentration.

Preparation of Tomato Paste and Tomato Puree

Ingredients

Materials	**Quantity**
Tomato juice or pulp	1 kg
Sodium benzoate	0.5 g

Procedure

Following steps are taken for preparation of tomato paste and tomato puree.

1. Extract tomato juice as given above with tomato juice
2. Take tomato juice into stainless steel pan or open cooker or cooking vessel
3. Boil it to desired consistency
4. Add preservative while boiling
5. Test the end point as given as above
6. If it is ready fill in bottles or jars
7. Put cap and tied it properly
8. Label bottle or jar
9. Store jar in dry place

5. Tomato Soup

Tomato soup is becoming very popular in homes. It can be stored in refrigerator and is warmed at the time of serving.

Preparation of Tomato Soup

Ingredients

Materials	**Quantity**
Tomato pulp or juice	1 kg,
Salt	20 g
Sugar	20 g
Butter or cream	20 g
Flour/starch	10 g
Onion (chopped)	20 g
Garlic (chopped)	5 g
Clove (headless)	5 numbers
Cumin	1 g
Cardamom (large)	1 g
Black pepper	1 g
Cinnamon (powdered)	1 g
Water	350 ml

Procedure

Steps taken are given below.

1. Select firm and fully ripe fruits
2. Wash fruits in clean and fresh water thoroughly
3. Extract tomato juice as given for juice preparation
4. Strain pulp through muslin cloth to remove seeds and skin
5. Take fruit juice into pan
6. Keep pan on regulated electric heater or Gas stove for boiling
7. Cook to desirable consistency
8. Take all spices in cloth bag and put in cooking juice
9. Continue juice boiling
10. Prepare paste of onion and garlic and add to concentrated juice
11. Prepare paste of butter and flour and add to boiling juice
12. Add sugar and salt to boiling juice
13. Remove spices bag and squeeze it
14. Boil it again for few minutes
15. Stop boiling
16. Fill concentrated juice into can or bottles
17. Sterilize can at 110 °C.
18. Cool cans at room temperature
19. Label can or bottles
20. Store them in dry and cool place or refrigerator
21. Warm soup when it is served

Flow Sheet for Preparation of Tomato Soup

Tomato pulp/juice

↓

Boil to concentrate it

↓

Take spices into a cloth bag

↓

Put spices bag into the boiling pulp for 5 minutes

↓

Prepare paste of butter and flour/starch

↓

Add to concentrated juice

↓

Cook mixture to desired consistency with continuous stirring

↓

Add sugar and salt to it, cook for 2 minutes

↓

Remove spices bag and squeeze it

↓

Fill into can and sterilize at 110 °C. for 40-45 minutes

↓

Cool and label can

↓

Store them

6. Tomato Cocktail

Preparation of Tomato Cocktail

It is prepared just before serving or served from stock. For stock preparation following quantity of ingredients are required.

Ingredients

Materials	**Quantity**
Tomato juice	5 litres
Salt	45 g
Lime juice	60 ml
Red Chilli powder	0.25 g
Clove (headless)	5 numbers
Cumin	1.5 g
Black pepper	1.5 g
Coriander seed	1.5 g
Cardamom (large)	1.5 g
Cinnamon (powdered)	1.5 g

Procedure

Steps mentioned below are taken for preparation of tomato cocktail.

1. Select uniformly ripe healthy fruits
2. Remove peduncle and calyx from the fruits
3. Wash them with clean and fresh water
4. Cut fruit into pieces
5. Press tomato pieces by wooden ladle or stainless steel spoon or knife to extract juice
6. Strain juice through a Muslim cloth piece
7. Take juice into stainless steel vessel
8. Put spices bag into juice and immerge it.
9. Boil juice for 20 minutes
10. Add lime juice, Vinegar, Salt and little sugar if needed
11. Heat mixture at 85 °C.
12. Remove spice bag and squeeze it into juice
13. Fill juice into sterilized bottles
14. Cork them by crown cork
15. Pasteurize them in hot water at 85 °C
16. Cool bottles
16. Label them
18. Store bottles in cool and dry place

Flow Sheet for Preparation of Tomato Cocktail

Uniformly ripe fruit

↓

Trimming

↓

Washing

↓

Extract juice

↓

Pouring juice into vessel

↓

Putting spices cloth bag into juice

↓

Boiling for 20 minutes

↓

Adding lime juice, vinegar and salt if needed little sugar

↓

Heating at 85°C.

↓

Filling into clean sterilized bottles

↓

Corking bottles

↓

Pasteurizing at 75-80°C

↓

Cooling

↓

Labeling

↓

Storage

7. Tomato Chilli Sauce

It is highly spiced product made from ripe, peeled and crushed tomatoes and salt, sugar, spices, vinegar, with or without onion and garlic. In this case Chilli powder is added more according to taste and pungency. The method of preparation is similar to that for tomato sauce.

Chilli Products

I. Chilli Pickles

Chilli pickles are of three types as given below

1. **Chilli and lime pickle:** It is prepared by mixing both fruits ofchilli and lime juice.
2. **Chilli pickle in salt:** This Chilli pickle is prepared for short or immediate use only.
3. **Large fruited chilli pickle in oil:** This type of pickle is prepared from large fruited variety of chilli in oil for long storage and future use. It is also very popular in some parts of country. Generally it is very commonly used in rural parts of Uttar Pradesh and Bihar.

1. Preparation of Chilli And Lime Pickle

The preparation of this type pickle is already described earlier

2. Preparation of Chilli Pickle With Salt

Ingredients

Materials	**Quantity**
Chilli fruits	1 kg
Salt	100 g
Mustard oil	sufficient quantity for frying

Procedure

Following steps taken for this type of pickle preparation. It is very popular in villages and prepared for immediate or short used.

1. Healthy green Chilli fruits are taken
2. They are washed with fresh and clean water
3. A slight longitudinal cut is given with the help a sharp stainless steel knife to remove seeds
4. Seeds are removed
5. Salt is filled inside the fruits
6. Fruits are fried in mustard oil
7. Fried fruits are filled in a clean jar or bottle
8. Keep for one day. It will become ready for use within a Day.

3. Preparation of Large Fruited Red Ripe Chilli Pickle in Oil

Ingredients

Materials	Quantity
Red ripe Chilli fruits	1 kg
Amchur	200 g
Salt	50 g
Fennel powder	10 g
Fenugreek powder	10 g
Nigella (Kalaunji)	10 g
Cumin powder	10 g
Mustard powder	10 g
Mustard Oil	350 ml (sufficient quantity to get fruits immersed)

Procedure

Following steps are taken.

1. Select healthy uniformly red ripe large fruits
2. Wash them with clean fresh water
3. Spread on cloth under shade to dry out moisture from skin surface
4. Trim flower stalk (pedicle) from the fruits
5. Remove seeds through the whole left by pedicle gently with the help of stainless steel knife. Fruits will become hollow.
6. Prepare mixture of spices by mixing them together.
7. Roast spice mixture and grind coarsely
8. Mix salt and Amchur with spice mixture
9. Fill this mixture inside the fruit through whole with the help of stainless steel spoon or a thin wooden stick
10. Press mixture inside the fruits gently by spoon or a wooden stick to remove air from the spice mixture
11. Put filled chilli fruits in a well clean and sterilized jar or wide mouth bottle.
12. Pour sufficient mustard oil so that fruits are immerged completely
13. Keep jar for few weeks so that pickle become ready for use.
14. Label the jar
15. Store in dry place

II. Chilli Sauce

It is already given in tomato and Chilli sauce preparation.

III. Chilli Paste

Fresh green or red fruits are for preparation of chilli paste. Chilli paste is prepared in the same manner of preparation of garlic and ginger paste. Both green and dry chillies are also available in liquid or paste forms

IV. Dried Red Chilli Fruits

Red ripe chilli fruits are dried in the sun and packed and preserved for future use. This practice is usually is followed in rural areas. Drying of chilli has been described with drying of vegetables section.

V. Chilli Powder

Dried chilli fruits are crushed or grinded manually or mechanically with help of machine. Chilli powder is packaged and stored for future use.

VI. Blends of Dry Chilli Powder of Various Spices

Chilli powder is also used in various spice blends such as curry powder e.g. Chana Masala, Chat Masala, Meat Masala, Chiken Masala.

In real sense no dish neither vegetarian and non-vegetarian is complete without chilli, especially in India. Chilli is not only commonly used in India but in Ethiopia and Eritrea, a blended chilli powder in various spices is very commonly used in every house for preparation of meat curry known as "Berbere". They do not cook meat without Berbere which provides a delicious taste to meat curry.

Carrot Products

Carrot is used for many preparations such as carrot juice, carrot halwa, carrot pickle with radish and turnip and carrot preserve.

I. Carrot Pickle

Carrot pickle is prepared by mixing with cauliflower, Turnip and Radish pieces in mustard oil by same procedure as given with cauliflower pickle. Here procedure for mixed carrot pickle with radish and turnip is given.

Preparation of Mixed Carrot Pickle

Ingredients

Materials	**Quantity**
Carrot	400 g
Radish	300 g
Turnip	300 g
Salt	50 g

Mustard oil	350 ml
Fennel	10 g
Fenugreek	10 g
Nigella (Kalaunji)	10 g
Cumin	10 g
Coriander powder	10 g
Asafoetida	2 g
Clove	5 Number
Black pepper powder	10 g

Procedure

Steps taken are as follows

1. Select healthy edible roots of carrot, radish and turnip
2. Remove stem and thin part of the roots with stainless steel knife
3. Wash roots with fresh and clean water thoroughly.
4. Remove small side roots
5. Peel with stainless steel knife or peeler
6. Cut roots into long, thick slices or pieces
7. Blanch them in boiling water at 85°C for 5 minutes.
8. Dry pieces in the sun for a day
9. Roast pieces slowly in a pan
10. Mix all spices with root pieces thoroughly
11. Fill in clean jar or bottles
12. Pour sufficient mustard oil so that pieces are immersed in oil.
13. Keep jar or bottle for few days till it is ready for use
14. Label the jar
15. Store pickle in dry and cool place.

II. Carrot Juice

Carrot juice is very popularly consumed in western parts of the country. Thoroughly cleaned and washed carrot roots crushed to extract juice by carrot juice extracting device or extractor. Juice extractor may be handle manually or electricity.

Preparation of Carrot Juice

Ingredients

Materials	**Quantity**
Carrot roots juice	1 kg
Black salt	According to taste (if needed)
Lime juice	5 ml (if needed)

Procedure

Following steps are taken for preparation for carrot juice.

1. Select healthy edible roots
2. Remove stem and thin part of the roots with stainless steel knife
3. Wash roots with fresh and clean water thoroughly.
4. Remove small side roots
5. Peel with stainless steel knife
6. Cut roots into long, thick slices
7. Extract juice with the help of juicer
8. Strain juice through a muslin cloth
9. Add or mix black salt and lime juice if needed according to taste
10. Add ice cubes to cool the juice
11. Use immediately

III. Carrot Preserve

Carrot roots can be used for preparation of preserve.

Preparation of Carrot Preserve

Ingredients

Materials	**Quantity**
Carrot roots	1 kg
Sugar	1 kg
Citric acid or Lime juice	5 ml
Water	100 ml

Procedure

The carrot preserve is prepared in the same manner as procedure given for mango preserve. The following steps are taken.

1. Select healthy edible roots
2. Remove stem and thin part of the roots with stainless steel knife
3. Wash roots with fresh and clean water thoroughly.
4. Remove small side roots
5. Peel roots with stainless steel knife or peeler
6. Cut roots into long, thick slices or pieces
7. Fork them with stainless steel fork
8. Blanch them in boiling water at 85°C for 5 minutes
9. Take a wide stainless steel bowl and equal quantity of sugar to the weight of fruits (1 kg fruit + 1 kg sugar).

10. Put a layer of sugar first then second layer of fruit pieces like wise put sugar and fruit pieces in layers one by one. Keep for a day
11. During this period root juice from the pieces will ooze out and dissolve sugar. Sugar will enter into root pieces. The whole sugar will become in solution form.
12. Add citric acid in required quantity or lime juice to the mixture of root pieces and sugar.
13. Keep on regulated electric heater or Gas for boiling. Let it boil for 10-15 minutes.
14. No need to add preservative because mixture contains more than 70 percent sugar which will not allow microbes to survive in the mixture.
15. When sugar syrup become thick consistency check the end point whether it is ready or not.
16. For this take a drop of syrup in a tea spoon and cool it
17. Take this drop between your finger and thumb and separate gently both. If it is ready a thread like structure will be formed, if not continue boiling till it is observed.
18. After observing end point close the gas and remove bowl from the Gas.
19. Keep it for cooling, now it is ready.
20. Fill the ready preserve or muraba in well cleaned wide mouth glass bottles.
21. Label jar or bottle
22. Store bottles at cool and dry place for future use.

Flow Sheet for Preparation of Carroot Muraba

Carrot roots

↓

Trimming

↓

Washing

↓

Peeling

↓

Cutting into slices or pieces

↓

Forking slices

↓

Putting carrot slices and sugar in alternate layers

↓

Keep stand for over night

↓

Drain sugar syrup from slices

↓

Add little water into syrup

↓

Boil syrup and put slices into syrup

↓

Continue cooking till syrup becomes thick consistency

↓

Test end point

↓

If ready stop cooking

↓

Remove preserve from gas and cool it

↓

Fill into cleaned and sterilized jars

↓

Closing jar

↓

Capping and Cooling jars

↓
Labeling jars
↓
Storage

Cauliflower Product

I. Cauliflower Pickles

It has already been described earlier.

Preparation of cauliflower pickle with mixture of salt, oil, spices and vinegar

Ingredients

Materials	**Quantity**
Cauliflower (pieces)	1 kg,
Salt	150 g
Ginger (chopped)	25 g
Onion (chopped)	50 g
Garlic (chopped)	10 g
Red chilli powder	15 g
Turmeric,	15 g
Cinnamon	15 g
Black pepper	15 g
Cardamom (large)	15 g
Cumin	15 g
Aniseed (powdered)	15 g
Cloves (headless)	6 numbers
Tamarind pulp	50 g
Mustard (ground)	50 g
Vinegar	150 ml
Mustard oil	400 ml

Procedure

Following steps are taken.

1. Select firm creamy white healthy curd
2. Remove outer leaves and undesirable parts of the curd.
3. Wash curd with fresh and clean water to remove dust and dirt from the surface.
4. Cut curd into pieces by separating curd segment.
5. Blanch them in boiling water for 5 minute at 85°C. This will soften the tissues.

6. Dry these pieces in sun for a day
7. Mix ingredients and prepare pickle.

Potato Products

Potato is semi-perishable tuber vegetable therefore; post-production management in potato is important aspect. Under tropical and subtropical conditions, 40-50% losses occur due to poor handling and storage. Therefore, it is important to minimize post-harvest losses. Processing of potatoes is advantageous, because it makes storage easier due to reduction in bulkiness and increase in shelf-life. It adds value to potatoes and, therefore, gives better returns. From the consumer point of view, all processed products should have an attractive colour, acceptable texture and good flavour. Generally, use of high quality potatoes will help to obtain and maintain good quality of the processed products. Potatoes are processed into many types of products such as: (i) fried products such as wafers/chips etc; (ii) Dehydrated products such as dice etc; (iii) Frozen products such as French fries, patties, puffs, dice etc and (iv) Canned. The raw material requirements for different forms of products may differ.

Important considerations in Potato Processing

The quality of potato required for processing are:

1. Variety having long dormancy
2. Varieties having round or oval tubers with medium to large size with fleet eyes
3. Variety free from diseases and peeling losses.
4. The dry matter content of the tubers should be high for greater yield of dehydrated products.
5. The sugar and phenol content should be low and the tubers should be free from darkening after cooking.
6. Potato varieties with high specific gravity (or dry matter) are preferred for preparation of chips, French fries or dehydrated products, whereas those of lower specific gravity.are used for canning or for such other products where a firm piece is desired. Yield of chips and flour etc. is higher from high specific gravity potatoes. Besides high yields of the product, the uptake of fat or oil during frying is lower. However, potatoes of very high specific gravity may not be suitable for the production of French fries etc. French fries produced from potato with a specific gravity of 1.106 were too hard and biscuit like. It is also of the opinion that very high specific gravity potatoes may yield ‘hard chips’. Dry matter content (specific gravity) is vital in terms of yield of potato chips or wafers and the texture of potato chips. Specific gravity of potatoes can be determined by a number of methods e.g. use of i) brine solution of known specific

gravities (Burton 1989), ii) Potato Hydrometer, iii) Archimedes principle (Nissen 1967), iv) variable load hydrometer. If interested please concern literature for detail procedure of measurement.

Generally Potato Chips, French fry, and Potato Papad are prepared by potato tuber.

I. Potato Chips

Potato chips are very popular and commonly used in every house in urban population. Therefore, many companies have started preparation of potato chips. However, potato chips are prepared at home scale also as described below.

Preparation of Potato Chips

Ingredients

Quantities of different ingredients used depend up on the taste and consumption.

Materials	Quantity
Potato tubers	1 kg
Edible oil	Sufficient amount for frying
Salt	According to taste
Black pepper powder	According to taste

Procedure

Following steps taken are describe below.

1. Select healthy potato tubers
2. Wash tubers with clean and fresh water
3. Remove defective and small tubers
4. Peel tubers with stainless steel knife or peeler
5. Remove residual skin and discoloured area and green parts
6. Cut slices 1-1.5 mm thick with stainless steel knife or slicer
7. Dip slices in salty water to avoid browning
8. Drain out salty water
9. Keep slices in trays or clean cloth in single layer
10. Keep trays in open sunlight or solar drier for drying
11. Fry dried slices in refined oil. Some time dried slices are packed and fried at use time
12. Add salt and flavouring materials.
13. Cool slices
14. Pack them
15. Store them for future use

Some steps are shown in following photos taken during class.

Selected washed potato

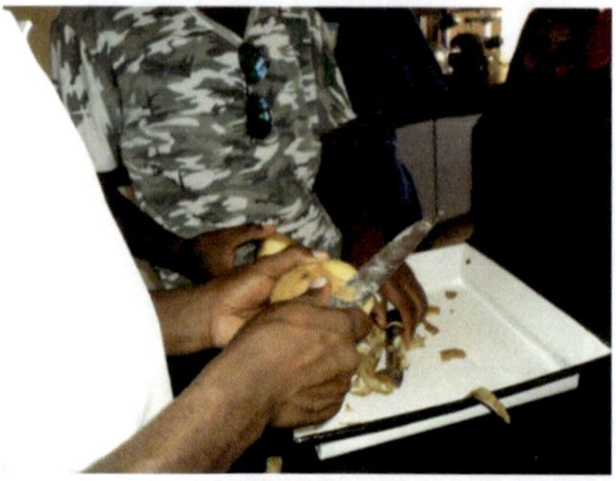

Peeling of potato

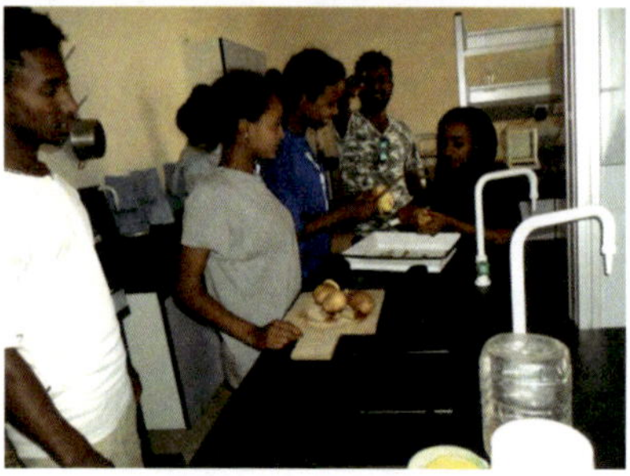

Treaming of poatato tuber

Ready peeled potato for slicing

Cutting slicing of peeled potato

Potato Slices

Slices ready for drying

II. Potato Finger Chips

Potato Finger chips are prepared for immediate use or consumption.

Preparation of Potato Finger Chips

Ingredients

Materials	**Quantity**
Potato tubers	1 kg
Mustard Oil	Sufficient amount for frying
Salt	According to taste
Black pepper powder	According to taste

Steps:

All steps are followed as given above. Only difference is potato long pieces are cut and fried for immediate use.

III. Potato Papad

Potato papad is not commonly popular as papd prepared from other gains.

Preparation of Potato Papad

Ingredients

Materials	**Quantity**
Potato (medium sized)	1 kg
Salt	Less than 1 tea spoon (according to the taste)
Red Chilli Powder	Less than a 1/2 tea spoon.
Oil	2 table spoon

Procedure

Following steps are followed or preparation of potato papad.

1. Select medium size of healthy potato tubers
2. Wash them with clean and fresh water
3. Boil potato until they get soft or tender
4. Cool down potatoes
5. Peel and remove skin
6. Grate or smash them
7. Add salt and red chilli powder into grated potato and mix thoroughly.
8. Smear oil on hand palm
9. Prepare balls of potato mixture and keep them in a stainless steel plate. About 20-22 balls can be prepared by 1 kg potato mixture.
10. Role out balls like chapatti with the help of chapatti maker (Chakal and Belan)

11. Keep papad on a wide big plate
12. Put papad on a piece of clean piece of cloth in sunlight for drying
13. Cover papad by muslin cloth to avoid dust.
14. After drying pack them
15. Label papad
16. Store in dry place

Flow Sheet for Preparation of Potato Papad

Selection of medium size tuber

↓

Washing

↓

Boiling

↓

Peeling

↓

Grating or smashing

↓

Mixing salt and red chilli powder

↓

Smashing or mixing thoroughly into smashed potato

↓

Making balls

↓

Rolling like chapatti

↓

Keeping on a piece of clean cloth in the sun for drying

↓

Covering by muslin cloth to avoid dust

↓

Drying

↓

Packaging

↓

Labeling

↓

Storage

Sweet Potato Products

I. Sweet Potato Chips

Sweet potato Chips are prepared as procedure given for potato chips preparation

II. Sweet Potato Finger Chips

This is also prepared in similar way as potato finger chips are prepared.

Onion Products

1. Onion Pickle

Onion is commonly used for making onion vinegar pickle. Generally small bulbs of onions are used for making pickle in vinegar.

Preparation of Onion Vinegar Pickle

Ingredients

Materials	**Quantity**
Onion	1 kg
Salt	250 g
Vinegar	1000 ml
Red chilli powder	10 g
Cumin	10 g
Cardamom	10 g
Clove	5 number
Black pepper	10 g

Procedure

Following steps are taken for preparation of onion vinegar pickle.

1. Select small healthy onion bulbs
2. Peel or remove outer dry skin covering bulb
3. Boil them in small quantity of water at 85°C. for 5 minutes
4. Drain out water from onion
5. Put in jar or wide mouth bottles
6. Add salt into bulbs
7. Keep standing for 24 or 48 hrs
8. Drain out water from onion
9. Pour vinegar and spices into onion bulbs
10. Mix them thoroughly
11. Seal jar
12. Label jar
13. Store jar in dry place

Flow Sheet for Preparation of Onion Pickle

Selection of small onion bulbs

↓

Peeling

↓

Blanching

↓

Draining water

↓

Filling blanched onion into jar

↓

Adding salt

↓

Keeping for 24 or 48 hrs

↓

Draining water from onion and salt mixture

↓

Adding vinegar and spices to onion mixture

↓

Labeling

↓

Storage

Ginger Products

Ginger is used for making ginger powder and ginger and garlic paste.

Ginger powder: Ginger powder is a grinded powder of dried ginger which is used for vegetable cooking, meat and chicken cooking and tea preparation.

Ginger and Garlic paste: This is for making vegetables soups, meat and chicken soup.

1. Ginger Powder

Preparation of Ginger Powder

Ingredients

Material	**Quantity**
Ginger rhizomes	1 kg

Procedure:

Following steps are followed for preparation of ginger powder:

1. Select good quality of ginger rhizomes
2. Wash with fresh and clean water
3. Scrape outer skin of the rhizomes
4. Wash them again with fresh water
5. Boil rhizomes in hot water at 85 °C. for 4 minutes
6. Remove rhizomes from water
7. Dry them in the sun
8. After drying grind rhizomes with grinder
9. Sieve grinded ginger
10. Pack ginger powder in good quality plastic bag or fill in the well cleaned bottles

11. Label bottles
12. Store bottles in dry place

Flow Sheet for Preparation of Ginger Powder

Ginger

↓

Washing

↓

Scrapping

↓

Washing

↓

Blanching

↓

Draining out water

↓

Drying in sun

↓

Crushing or grinding to make powder

↓

Filling in bottles or Packing in packing bags

↓

Labeling

↓

Storage

2. Ginger Garlic Paste

Preparation of Ginger And Garlic Paste

Ingredients

Materials	Quantity
Ginger	500 g
Garlic	500 g
Chilli	10 g

Procedure

Following steps are followed for preparation of ginger + garlic paste.

1. Select healthy Ginger, Garlic and Chilli fruits
2. Wash ginger and chilli separately with fresh and clean water
3. Peel garlic cloves
4. Crush ginger, chilli and garlic cloves separately by pastel and marter to make paste
5. Take pastes of ginger, chilli and garlic into a stainless steel vessel or pan and mix together thoroughly
6. Boil mixture to temperature at 85 °C. for 5 minutes till sufficient Brix 33°is obtained.
7. Fill paste into clean sterilized bottles
8. Cool bottles at room temperature
9. Label bottles
10. Store in dry place.

Flow Sheet for Preparation of Garlic and Ginger Paste

Garlic, Ginger and Chilli

↓

Washing

↓

Trimming

↓

Scraping or peeling

↓

Washing with chlorinated water and then fresh normal water

↓

Cutting garlic, ginger and chilli into pieces

↓

Mixing them

↓

Crushing or grinding to make paste

↓

Mixing paste thoroughly

↓

Heating paste at 85°C. for 5 minutes

↓

Fill paste in bottles

↓

Cooling

↓

Label bottle

↓

Store

Drying of Vegetables

Some vegetables can be dried and preserve for future and consumed during off season or scarcity period. Usually they are dried in open sunlight on home scale. Vegetables commonly dried on home scale in rural areas are:

1. Cauliflower,
2. Fenugreek leaves,
3. Chana sag,
4. Kasuri methi
5. Okra.
6. Red ripe chilli fruits

Different steps for drying are described as follows:

Drying Process for Different Vegetables

Procedure

Vegetables are dried in this way.

1. Selection of healthy vegetables
2. Undesirable parts or diseased portion are removed.
3. They are washed in fresh and clean lean water
4. They are cut in to pieces or slices.
5. Chopped pieces or slices are blanched (heated) for 5 minutes in boiling water at 85°C. This will soften tissues.

   The main point is how to maintain natural colour in dried vegetables. Therefore, at the time blanching boiling water should contain the following salts which are recommended to retain maximum natural colour in dried form. In some cases sulphuring is done for this purpose. Sometime vegetables are dried under semi- shady places to retain natural colour in dried vegetables. Generally it is followed in rural areas.

   To retain natural colour in dried vegetables boiling water should contain the following quantity of salts used in blanching water.

Salts	Quantity
Magnesium oxide	0.1 percent
Sodium carbonate	0.1 percent
Potassium metabisulphite (KMS)	0.5 percent

6. After that they are spread on the cloth in thin layer and covered with a thin muslin cloth to avoid dust contamination. Heat treated vegetable pieces or slices are dried in the open sun for 4-5 days.
7 Well dried vegetables are packed or filled in well cleaned dried jar or bottles
8. Jars are labeled
9. Jars are store in dry place

Drying of Cauliflower

Cauliflower is dried traditionally in villages. Curd is cut into piece and dried in the sun, packed and stored for future use.

Procedure

1. **Selection:** Select curds having milky white or creamy white colour and firm or solid. Ricy curd and curds having browning should be discarded.
2. **Washing:** Wash curds in clean and fresh water thoroughly.
3. **Sorting:** Remove bruised and undesirable portion of curd.

4. **Cutting:** Separate segments of curd with help of stainless steel knife. If segments are large in size the cut segments longitudinally into two halves.
5. **Blanching:** Segments of curds are blanched (heated) for 5 minutes in boiling water at 85°C. This will soften tissues.
6. **Draining out water:** Drain out water from segments.
7. **Drying in the sun:** Spread blanched segments in a thin layer on cloth sheet and kept on open floor where sufficient sunlight is available for the whole day. It will become dry within 8-15 days. Segments can also be dried in Solar dryer.
8. **Packaging:** Dried cauliflower curd segments are filled in containers
9. **Capping:** Put cap and fix tightly so that dried materials should not absorbe water vapour from the atmosphere.
10. **Labeling and storage:** Containers are labeled and stored in dry place.

Care should be taken while drying so that no animal and bird will finish the materials.

Drying of Chilli

India is homeland of spices and more than 60 spice crops are grown in the country and several spice crops are originated from India. Since ancient time country is famous for spice trade. Black pepper had been most important spice crop in the Indian trade and is known as king of spices. In deed there is confusion between black pepper and hot pepper to some people. There is a very interesting story about the drying of red chilli in the world when Drying of chilli has played very important role in the discovery of New World. When Columbus was on the way to India to get Black pepper, he missed his way and reached to America. While landing on the sea shore he saw people were drying red chilli fruits. He became very glad and called them Red Indian because he was narrated about the Indian feature, those people resemble with Indian. He filled his ships with red dried chilli and returned back but he failed to reach India to collect Black pepper.

To obtained dry chilli fruits, fruits are not harvested before ripening and left intact with plant in field. When fruits are fully ripe and become red in colour are harvested. Fully red ripe fruits are picked at an interval of 1-2 weeks and harvesting continues for a period of about three months.The yield of dried fruit varies with chilli crop growing conditions. The average yield of dry chilli fruits obtained under rainfed condition is 10-15 q/ha and under irrigated conditions 20-25 q/ha. Harvested ripe fruits are spread in thin layer on threshing floor or on cloth sheet for sun drying. Care is taken that the natural fruit colour should not faded by too long sun drying. Traditionally ripe chillies fruits are dried under sun for 8-15 days, while commercially it is dried at about 54.4 oC in 2-3 days. Uniform size (length-1.5 -2.0 inches) bright red colour without white chilli are selected for drying. Well dried Chilli fruits have 12% moisture. Traditionally red ripe fruits can be dried in the sun on open floor or fruits are tied with a thick thread or thin rope and hanged in the sun as given below.

Drying chilli on floor

Drying chilli by hanging method

Procedure for drying chilli

1. Selection of fruits: Well red ripe, healthy and uniform size fruits are selected.
2. Spread fruits in thin layer on floor or cloth sheet
3. Cover fruits with a thin muslin cloth to avoid dirt contamination of dust.
4. Some time change the sides of fruits by hand or with the help of a stick. Use of a stick would be better to burning effects of chilli fruits in hands.
5. Red ripe fruits will become dry within 8-15 days and have 12% moisture content.
6. Well dried fruits are cleaned and packed in packaging containers.
7. Label the packages
8. Store in dry place.

Red ripe chilli fruits can be dried in Solar dryer also on small scale. Drying of fruits in solar dryer has been explained in physical method of processing.

Drying of Leafy Vegetables

Leafy vegetables such as Fenugreek both Desi methi and Kasuri methi and Chana sag are dried open sun light and stored for future consumption.

Procedure

Following steps are followed for drying of leafy vegetables.

1. Healthy twigs with leaves are selected. All diseased, yellowish coloured spotted or damaged one are discarded.
2. They are washed thoroughly with fresh and clean water
3. After washing they are spread in a thin layer to remove and dry moisture.
4. Materials are cut into small pieces with stainless steel knife
5. Pieces are spread in a thin layer on cloth put on the floor in the sunlight for drying.
6. It is covered by a thin muslin cloth to avoid dust contamination
7. It will take 4-5 days to get dried in sunlight. Care should be taken if it is kept for drying in bright sunlight proper green colour could not be maintained. Therefore always semi sunlight is preferred.
8. When pieces are fully dried they are filled in clean wide mouth bottles or jars.
9. Jars are labeled
10. They are stored at room temperature.

Flow Sheet for Drying Vegetables

Vegetable

↓

Washing

↓

Trimming

↓

Cutting or slicing

↓

Blanching

↓

Drying

↓

Filling in bottles or jar or packaging

↓

Labeling

↓

Storage

16

Packaging and Packaging Materials

Packaging

Packaging is a mean or system by which a fresh or processed product will reach from the production center to consumer in safe and sound condition at an affordable price.

Objectives of packaging and packaging materials

The primary objective or packaging is to protect the contents during storage, transportation and distribution against deterioration. It may be physical, chemical or biological. However, followings are objectives of packaging and packaging materials.

1. To prevent loss through leakage, spillage, or pilferage
2. To protect contents from external physical, mechanical, and biological forces throughout storage, transport, and marketing
3. To preserve contents and prevent or retard direct and indirect chemical decomposition or diminution of contained-product quality
4. To facilitate ease of filling and closure and provide satisfactory closure integrity

Advantages of packaging

1. It provides effective protection to package food against external environments
2. It preserves the quality of food.
3. It is one of the most important parameter that sale products
4. It also provides detail information about the product like, content, ingredients, nutritional values
5. It also provides packaging and expiry dates.
6. it is useful in creating conditions that extend the storage life of a food.
7. It provides sanitary protection
8. It provides gas and odoour protection
9. It provides light protection
10. It provides transparency

Packaging Materials and Containers

There are several types of packaging materials and containers available for packing of fresh as well as processed fruits and vegetables products. Packaging fresh fruits and vegetables is one of the most important steps in the long and complicated journey from grower to consumer. Fresh fruits and vegetables are generally packed in bags, crates, baskets, cartons, bulk bins and pelletized containers which are convenient for handling, transport and marketing of fresh produce. Processed products are packed in metal containers or glass jars because they maintain their nutritional and organoleptic quality for extended periods of time due to durability and excellent oxygen and moisture barrier properties of the containers. Plastic packaging materials have recently become more popular due to their light weight and unbreakable nature and convenience features. Plastic polyethylene terephthalate (PET) containers are also becoming popular for high acidic fruits products.

Characteristics of Packaging Materials

1. They must be non-toxic and compatible with specific food
2. They should be lighter in weight
3. They should be easier to open
4. They should be Easier to reseal and store
5. They should ensure safe transport and distribution
6. They should protect the product from adverse effects such as heat, cold, moisture etc.
7. Packaging materials should have sufficient strength and rigidity so that damage can be reduced due to compression load.
8. They should be resistant to impact
9. They should have temperproofness
10. They should have good pouring features
11. They should be ease of disposal
12. They should not have size, shape, and weight limitations
13. They should have good appearance and printability
14. They should have low cost.

Classification of Packaging Materials

Packaging materials commonly used for foods may be classified as:

1. **Flexible** - paper, thin laminates, and plastic film
2. **Semi-rigid** - aluminum foil, laminates, paperboard, and thermoformed plastic
3. **Rigid** - metal, glass, and thick plastic

Based on nature of produce their packaging containers can be grouped in to two groups primary containers and secondary containers.

Primary packaging containers: The primary containers are those which come in direct contact with the food or products. Some foods and products such as fruit concentrates are filled in highly impervious to water vapour and other gases containers are known as primary packaging containers. Metal, glass and rigid plastic materials may be used for primary or consumer packages.

Secondary packaging containers: Somme products are provided with natural primary containers by nature such as nuts, orange etc. They need only secondary packaging containers like Carton, outer box, wrap and drum to hold units together and give gross protection.

Hermetic closure: Two conditions of the greatest significance in packaging are hermetic and non-hermetic closure.

Hermetic containers: The term hermetic means a container which is absolutely impermeable to entry of gases and vapour throughout its entry, including its seams. Such containers as long as it remains intact, will automatically be impervious to bacteria, yeasts, moulds and dirt from dust and other sources since all of these agents are considerably larger than gas or water vapour molecules. The most common hermetic containers are rigid metal cans and glass bottles although faulty closures can make them non- hermetic. With very rare exceptions flexible packages are not truly hermetic for one or more of the following reasons. First, the thin flexible films, even when they do not contain minute pinholes, generally are not completely gas and water-vapour impermeable although the rates of gas and water vapour transfer may be exceptionally slow; second, the seals are generally good but imperfect; and third, even where film materials may be gas- and water-vapour-tight, such as certain gages of aluminium foil, flexing of packages and pouches leads to minute pinholes and crease holes. Glass containers are hermetic provided that the lids are tight.

Non hermetic containers: A container which prevents entry of micro-organisms is known as on-hermetic container.

Types of Packaging Materials and Containers

1. Films and Foils, Plastics

Films and Foils have different values for moisture and gas permeability, strength, elasticity, inflammability and resistance to insect penetration. Many of these characteristics depend upon the film's thickness. The important characteristics of the types of films and foils commonly used in food packaging are in table 1. For the most part such films are used in the construction of inner containers. Since they are non-rigid, their main functions are to contain the product and protect it from contact with air or water vapour. Their capacity to protect against mechanical damage is limited, particularly when thin films are considered.

Table 1: Characteristics of different types of films and foils

Material	Properties
Paper	Strength; rigidity; opacity; printability.
Aluminium foil	Negligible permeability to water-vapour, gases and odours; grease proof, opacity and brilliant appearance; dimensional stability; dead folding characteristics.
Cellulose film (coated)	Strength; attractive appearance; low permeability to water vapour (depending on the type of coating used), gases, odours and greases; printability.)
Polythene	Durability; heat-sealability; low permeability to water-vapour; good chemical resistance; good low-temperature performance.
Rubber hydrochloride	Heat-sealability; low permeability to water vapour, gases, odours and greases; chemical resistance.
Cellulose acetate	Strength; rigidity; glossy appearance; printability; dimensional stability.
Vinylidene chloride	Low permeability to water vapour, gases, copolymer odours and greases; chemical resistance; heat-sealability.
Polyvinyl chloride	Resistance to chemicals, oils and greases; heat-sealability.
Polyethylene terephtalate	Strength; durability; dimensional stability; low permeability to gases, odours and greases.

Source: FAD/WFP, (1970) (http://www.fao.org/3/V5030E/V5030E0h.htm)

These materials can exist in many forms, depending upon such variables as identity and mixture of polymers, degree of polymerisation and molecular weight, spatial polymer orientation, use of plasticisers (softeners) and other chemicals, methods of forming such as casting, extrusion or calendaring, etc

2. Plastic Sheets

Plastic materials are widely used in food packaging because they are relatively cheap, lightweight, and easy to form into desired shapes.

(a) Cellophane paper can be used for packaging of dried products mainly candies and fruit leathers.

(b) Polyethylene sheets have a variety of uses. They are flexible, transparent and perfectly resistant to low temperature and impermeable to water vapour.

Advantages

1. It can be easily heat sealed.
2. It can be utilized in the forms of sheets and bags.
3. It is a good packaging material for primary protection of dehydrated products.
4. When it is combine with other packaging materials, it provides good protection if needed to prevent flavor and gas losses.

3. Glass Containers

The glass container has its own importance for food packaging but usually have problems of corrosion and reactivity of metal closures and limitation of its susceptibility to breakage which may be from internal pressure, impact or thermal shock that can be minimize by proper matching of container to its intended use and intelligent handling practice. However, the heavier jar is more susceptible to both thermal shock and impact of breakage. Greater thermal shock breakage of jar is due to wider temperature differences which cause uneven stress between the outer and inner surfaces of the thicker glass. Thicker wall of heavier jar has greater impact breakage susceptibility due to lower resilience of its thicker wall. Thermal shock can be minimize with good practice to minimize temperature differences between the inside and outside of glass container. Some manufacturers recommend that a temperature difference of 44°C between inside and outside not be exceeded.

Classification of glass containers

Glass used for receptacles in fruit and vegetable processing is a carefully controlled mixture of sand, soda ash, limestone and other materials made molten by heating to about 1500°C (2800°F).

Main classes of glass receptacles are

(a) Jars which are resistant to heat treatments,

(b) Jars, glasses, etc. for products not submitted to heat treatment (marmalades, acidified vegetables, etc.);

(c) Glass bottles for pasteurized products (tomato juice, fruit juices, etc.. or not pasteurized (syrups. and

(d) Receptacles with higher capacity (flasks, etc.)

(a) Jars for sterilised/pasteurized canned products

Considering both the advantages and disadvantages these receptacles may replace metal cans.

Advantages

1. They do not react to food content.
2. They are transparent and can be manufactured in various shapes.
3. They use cheap raw materials and are reusable.

Disadvantages

1. They are heavier than metal can of same capacity;
2. They are fragile.
3. They have lower thermal conductance and a limited resistance to thermal shocks.

Receptacles in this category must assure a perfect hermeticity after their pasteurization/sterilisation and cooling and this has to be achieved by the use of metallic (or glass) caps and specific materials for tightness.

Taking into account the receptacles' closure method, there are two categories of receptacles:

1. Glass jars with mechanical closure;
2. Glass jars with pneumatic closure;

(b) Jars for products without heat treatment

Glass jars with non hermetic closures using metal, glass or rigid plastic caps are used for marmalades, jellies and jams, however for these products the receptacles mentioned above may also be used.

(c) Glass bottles

These receptacles are widely used both for finished products which need pasteurization (e.g., Tomato juice, Knit juices, etc.) and for those which are preserved as such (e.g., Fruit syrups).

Glass bottles in category (a) are closed hermetically with metallic caps, provided with special materials for tightness.

For glass bottles in category (b) various corks, and aluminium caps with tightness materials may be used.

Glass receptacles with high capacity: In this category there are glass flasks with 3 and 10 litre capacity which can be hermetically closed by a SKO caps system and are resistant to product pasteurization (ea. tomato juice).

4. Paper Packaging

It is a primary container and few paper products are not treated, coated or laminated to improve their protective properties. Paper from wood pulp and reprocessed waste paper will be bleached and coated or impregnated with such materials as waxes, resins, lacquers, plastics, and laminations of aluminium to improve water vapour and gas impermeability, flexibility, tear resistance, burst strength, wet strength, grease resistance, sealability, appearance, printability, etc.

(a) Paper sheets

Paper sheets can be categorised into following categories.

1. **Kraft paper:** It is the brown unbleached heavy duty paper commonly used for bags and for wrapping. It is seldom used as a primary container;
2. **Parchment paper:** It is made by acid treatment of paper pulp modifies the cellulose and gives water and oil resistance and considerable wet strength to this type of packaging material.
3. **Glassine-type papers:** They are characterised by long wood pulp fibres which impart increased physical strength.
4. **Paper with plastic material sheets.**

Receptacles from paper or cardboard (paper = 8 to 150 g/m^2; cardboard = 150 to 450 g/m^2).

5. Tin Can/Tinplate

The "tin can" is a container made of tinplate. Tinplate, a rigid and impervious material, consists of a thin sheet of low carbon steel coated on both sides with a very thin layer of tin. It can be produced by dipping sheets of mild steel in molten tin (hot-dipped tinplate) or by the electro-deposition of tin on the steel sheet (electrolytic tinplate). With the latter process it is possible to produce tinplate with a heavier coating of tin on one surface than the other (differentially coated). Tin is not completely resistant to corrosion but its rate of reaction with many food materials is considerably slower than that of steel. The effectiveness of a tin coating depends on:

1. Its thickness which may vary from about 0.5 to 2.0 µm (20 to 80 □ 10(-6) in.);
2. The uniformity of this thickness;
3. The method of applying the tin which today primarily involves electrolytic plating;
4. The composition of the underlying steel base plate;
5. The type of food, and
6. Other factors.

Some canned vegetables including tomato products actually owe their characteristics flavours to a small amount of dissolved tin, without which these products would have an unfamiliar taste. On the other hand, where tin reacts unfavourably with a particular food the tin itself may be lacquer coated.

The thickness of tinplate sheets may vary from 0.14 mm to 0.49 mm and is determined by weighing a sheet of known area and calculating the average thickness.

Tinplate sheets may be lacquered after fabrication to provide an internal or external coating to protect the metal surface from corrosion by the atmosphere or through reaction with the can contents. They may also be printed by lithography to provide suitable instructions or information on containers fabricated from tinplate sheets (otherwise paper labels can be attached to the outer tinplate surface).

Under normal conditions the presence of the tin coating provides a considerable degree of electrochemical protection against corrosion, despite the fact that in both types of tinplate the tin coating is discontinuous and minute areas of steel base plate are exposed. With prolonged exposure to humid conditions, however, corrosion may become a serious problem.

Common Organic Coating Materials

Common organic coatings of FDA approved materials and their uses are indicated in Table 2.

The coatings not only protect the metal from corrosion by food constituents but also protect the foods from metal contamination, which can produce a host colour and flavour reactions depending upon the specific food. Particularly common are dark coloured sulphides of iron and tin produced in low acid foods that liberate sulphur compounds when heat processed, and bleaching of red plant pigments in contact with unprotected steel, tin, and aluminium.

Table 2: General types of can coatings

Coating	Typical uses	Type
Fruit enamel	Dark coloured berries, cherries and other fruits requiring protection from metallic salts	Oleoresinous
C-enamel	Corn, peas and other sulphur-bearing products	Oleoresinous w. suspended zinc oxide
Citrus enamel	Citrus products and concentrates	Modified oleoresinous
Beverage can enamel	Vegetable juices; red fruit juices; highly corrosive fruits; non-carbonated beverages	Two-coated w. resinous base coat and vinyl top coat

Source: Ellis (1963) (http://www.fao.org/3/V5030E/V5030E0h.htm)

Criteria for Selection of Packaging Material for Processed Foods

Packaging provides quality and quantity assurance besides creating hygienic environment for food product. It offers security through tamper proof designs and contributes to the product image through structural and graphical design. Food safety is permanent importance as package products against biological, chemical and distribution damages. The selection criteria of packaging materials for processed food depend upon the following points.

1. **Mechanical damage:** Fresh, processed and manufactured foods are susceptible to mechanical damage. For examples are bruising of soft fruits and the break-up of heat processed vegetables. Such damage may result from sudden impacts or shocks during handling and transport, vibration during transport by road, rail and air. It may also occur due to compression loads imposed when packages are stacked in warehouses or large transport vehicles. Appropriate packaging can reduce the incidence and extent of such mechanical damage. Besides packaging, proper and good handling and transport procedures and equipment are also necessary.
2. **Permeability Characteristics:** In the case of packaging films, laminates and coated papers, the permeation of water vapour, gases like O_2, CO_2, N_2 and ethylene and volatile odour compounds into or out of package is an important consideration.

   Dried products are with relatively low moisture contents they will tend to pick up moisture, when they are exposed to a high humidity atmosphere. If the water activity of a dehydrated product is allowed to rise above a certain critical level, microbiological spoilage may occur. In such cases a packaging material with a low permeability to water vapor, effectively sealed, is required. A packaging material which is semipermeable to water vapor is required in this case. In such cases, the packaging material should have a low permeability to gases and be effectively sealed. To retain the pleasant odor associated with many foods, it is necessary to select a packaging material that is a good barrier to the volatile compounds which contribute to that odor. Such materials may also prevent the contents from developing taints due to the absorption of foreign odors.
3. **Temperature:** A package must be able to withstand the changes in temperature without any reduction in performance or undesirable change in appearance. This has importance when foods are heated or cooled in the package. Metal and glass containers were used for foods which were retorted in the package. Recently heat resistant laminates are developed for this purpose. are not suitable for packaging frozen foods.

   The method of heating may influence the choice of packaging. Many new packaging materials have been developed for foods which are to be processed or heated by microwaves. The rate of change of temperature may be important. For example, glass containers have to be heated and cooled slowly to avoid breakage.

4. **Light:** Many food components are sensitive to light, particularly at the blue and ultraviolet end of the spectrum. Vitamins may be destroyed, colors may fade and fats may develop rancidity when exposed to such light waves. The use of packaging materials which are opaque to light will prevent these changes. Pigmented plastic bottles are used for some health drinks.

Chemical Compatibility of The Packaging Material and the Contents of the Package

It is essential in food packaging that no health hazard to the consumer should arise as a result of toxic substances, present in the packaging material and leaching into the content. Two questions need to be answered to establish the safety of such packaging materials:

1. Are there any toxic substances present in the packaging material?
2. Will they leach into the product?

Testing of toxicological chemical compound is lengthy, complicated and expensive. It is usually involved extensive animal feeding trials and require expert interpretation of the results.

Protection against Microbial Contamination: The role of the package may be to prevent or limit the contamination of the contents by microorganisms from sources outside the package. In case of metal can some closures for glass containers are also effective barriers to contamination. The sealing requirements for containers for pasteurized products and foods preserved by drying, freezing, curing, etc. are not so rigorous. However, they should still provide a high level of protection against microbial contamination.

In-Package Microflora: The permeability of the packaging material to gases and the packaging procedure employed can influence the type of microorganisms that grow within the package. Highly permeable Packaging foods in materials to gases do not bring any significant change in the microflora, compared to unpackaged foods. There is a danger that pathogenic microorganisms could flourish under these conditions and result in food poisoning.

17

Research Needs for Product Development of Fruits and Vegetables

Several fruits and vegetables are unexploited for proper utilization and product development. Due to lack of utilization, there is a lot of wastage and ultimately the great loss to the people as well as nation. Some unexploited fruits and vegetables have potential for introducing new products. Therefore unexploited fruits and vegetables have to be exploited for new product development especially from underutilized sub-tropical and tropical fruits and vegetables. Research and development should be undertaken to utilize these fruits and vegetables.

Unexploited Fruits

1. **Unripe Green Mango:** Although pickles are prepared but droped green mango fruits are not preferred for pickle preparation. They can be dried and utilized for preparation of dried slices or amchur. Amchur is used to develop flavor and taste in Ice cream. These products are used to develop flavor and taste in meal or food. It is also used to prepare chatney. But utilization of Amchur is not known to common people.
2. **Avocado:** It is found in tropical and sub-tropical climatic regions having more humidity and rainfall. In Ethiopia it is found throughout the country but not utilized properly by people. They prefer to eat ripe fruits with salt and lime juice. It is very healthy fruit for chronic disease like cardiac and diabeted diseases. This is seasonal use but no product is available for off season. Therefore researches should be carried out for different products development of Avocado products.
3. **Jackfruit:** Tender fruits are cooked as vegetable whereas ripe fruits eaten in some parts of producing areas. Jackfruit is used for preparation of pickle in some houses which is negligible. No preserved products are found for off season.
4. **Kazmira:** A fruit found in Ethiopia specially in Harar areas, which highly perishable and starts spoiling after some times. Skin of fruit is very thin and soft which favours spoilage of fruits. Fruit are not used by people. Therefore, it needs to be processed to utilize this fruit properly.

5. **Cape Goose Berry:** It is a seasonal annual solaneous small fruit which is eaten raw by some people. Literature says that a product like jam can be prepared by this fruit but not popular in the society. Therefore it also needs to carry out some research on processing and product development.
6. **Karonda:** It is a also a small fruit used for making chatney and candy by some farming community people. But most of the people do not know about this fruits.
7. **Tind:** It is a wild small fruit found in desert areas which is used for making pickle only. Other products need to be exploited from this wild fruit.
8. **Mahua Flowers and Fruits:** Only ripe fruits are eaten by some people and unripe fruit is used for cooking vegetable which is negligible. Flowers of this fruit is sweet, fleshy and juicy which is for preparation of halwa and sweet puri. Dried flowers are used to prepare some snacks with safflower and linseed in villages during rainy season only. These are only seasonal uses of Mahua and no processed products are prepared by this valuable fruit for future use.
9. **Wild Ber:** Only dried fruits are used by children in some country like Eritrea. Some people prepare powder (Amchur) for future use. Wild ber fruits have sweet and sour taste so that ber powder is used like mango fruit dried powder. Wild ber fruits need to processed into other products.
10. **Phalsa:** Phalsa fruits have sweet and sour taste and liked by people. Ripe fruits are consumed as raw and used for making sharbat. No other processed products are available made by Phalsa.
11. **Aonla:** Aonla fruits have sour taste but after eating a person drinks water, sweetness is observed. It is very rich source of Vitamin 'C' and having high medicinal value. It is used only for making pickle and preserve (Murabba. on home scale. However, considering the importance of this fruit some companies have started processing of Aonla fruits and some processed products like candy and preserve are available.
12. **Karambola:** I t is eaten by some people as raw and there is no any product developed from Karambola fruits.
13: **Tamarind:** It is used only in preparation of some food to develop taste especially sour taste. Tamarind is very commonly used in southern dishes of country. No processed products are available.
14. **Fruit Waste:** (*i*) ***Banana skin:*** It is thrown after peeling banana fruits. No one knows its use but literature says it can be used for preparation of some products. (*ii*) ***Mango Skin:*** It is thrown after peeling mango fruits but it can be used for preparation of some products.

Vegetables

1. **Eggplant:** It is cooked as vegetable. There is no any product developed by this popular vegetable except pickle as literature says that pickle is prepared by brinjal fruits.
2. **Turnip:** There is no any product developed by this vegetable except pickle with radish
3. **Radish:** Only used for making pickle.
4. **Carrot:** No any preserved product is available of this nutritionally important root vegetable. However, carrot juice, carrot halwa and pickle are prepared. Carrot juice is common in some urban areas of Haryana and Delhi. Carrot halwa are available in Halwai shops in the market. Carrot halwa is also prepared in rural areas for home consumption. Carrot pickle is also prepared by rural people. There is no any large production of carrot processed products.
5. **Knol Khol:** Knol Khol is not a popular exotic vegetable in the country. It is only eaten as salad or cooked as vegetable. No any product is developed.
6. **Snap Melon:** Tender unripe fruits and ripe fruits are eaten and no any product has been developed.
7. **Bitter Gourd:** No processed product has been developed from this medicinally valuable vegetable. Fruits are cooked as vegetable and used for pickle preparation by some people.
8. **Bottle Gourd:** Fruits of his cucurbitaceous vegetable are only cooked as vegetable. No any processed product available.
9. **Pumpkin:** Pumpkin fruits are only cooked as vegetable. There is no any work has been done processing of this vegetable. However, author tried to prepare pumpkin candy successfully which has been describe earlier.
10. **Water Melon:** Only eaten as raw and rind is thrown but both fruits and rind of fruit can be utilized for preparing some processed products.
11. **Muskmelon:** Like watermelon it is also used as raw fruit which is very sweet and liked by people. But no processed product has been prepared.
12. **Wild Bitter Gourd:** Wild bitter gourd is found in the forest of some states. It is only cooked as vegetable. Wild bitter gourd fruits can used for processing when it is available during harvesting season.
13. **Pointed Gourd and Coccinia:** Fruits of both vegetables are cooked as vegetable only. Some Halwais prepare pointed gourd sweet only. No other processed products of both vegetables are available.
14. **Spine Gourg:** Spine gourd is also a wild cucurbitaceous vegetable found in the forest. I have seen small fruited spine gourd is found in

the forest of many states but large fruited spine gourd is found in the forest of Jharkhand state. I have both small fruited and large fruited spine gourds are sold in the Ranchi market. It is only cooked as vegetable but not popular.

15. **Sweet Gourd:** It is a wild cucurbitaceous vegetable cooked only as vegetable.
16. **French Bean:** It is cooked only as vegetable.
17. **Cluster Bean:** It is a popular legume crop of rainfed conditions. Industrially it is used for preparation of Guargum. In rural areas it used as cattle feed. Vegetable type variety has been developed and fruits are cooked as vegetable. There is any preserved product of this vegetable.
18. **Cowpea:** It is also a legume. There are many types of cowpea. Some of them having thick and fleshy and good quality of fruits. Fruits are cooked as vegetable. No processed product is available.
19. **Braod Bean:** Fruits are cooked as vegetable but not popular.
20. **Wild Beans:** There are some wild beans found in Ethiopia which are utilized only by those who are known about them.
21. **Sweet Potato:** It is a root tuber vegetable and commonly consumed in some African countries. Though it is not very common vegetable but it has high carbohydrate content. It plays very vital role in food nutrition security of any nation. No work on processing has been taken on this valuable vegetable crop.
22. **Black Carrot:** It is commonly grown only for cattle feed in some state like Punjab. Some people prepare Kanzi a type drink from black carrot in Punjab state of India. It needs to be processed on large scale and popularize and other processed products can be prepared.
23. **Cauliflower:** Curd is cooked as vegetable. In rural areas it is for preparation of pickle. Some time it is dried for future use.
24. **Leafy Vegetables:** Methi and Chana sags are dried and kept for future use in villages. Researches on processing of other leafy vegetables should be carried out to preserve these highly perishable vegetables.

A survey of the underutilized tropical fruits has also projected the tremendous potentials for introducing variety of new products of considerable commercial and nutritional status. Such tropical fruits are numerous such as:

Falsa (*Grewia asiatica*)
Karonda (*Carassa carandas*)
Jungli Badam or wild Almond (*Terminalia catapa*)
Jamun
Jaiful
Imbly or Tamarind (*Tamarindus indica*)
Dates
Melons and other cucurbits
Small purple berry

As per my knowledge there are several fruits and vegetables are underutilized crops and no attentions have been given on these crops for scientific exploitation.

Besides exploitation of unutilized fruits and vegetables, researches on these aspects should also be carried out for successful results.

There are few reports regarding maintenance of natural colour and blanching and sulpfuring time while drying process. Researches should also be carried out regarding the natural colour maintenance of dried fruits and vegetable products.

There is also great scope of research on blanching and sulfuring time of fruits and vegetables before drying.

There are few repots on packaging of processed fruits and vegetables, more research works are needed on this aspect for extension of shelf life of processed fruits and vegetables.

Useful Study Materials

1. Home-Scale Processing and Preservation of Fruits and Vegetables. Central Food Technological Research Institute, Mysore, India, Enlarged edition 1990.
2. Chavan, U. D. and J. V. Patil (2013) Industrial Processing of Fruits and Vegetables, Daya Publishing House, Astral International Pvt. Ltd. New Delhi-110002, 2013
3. Study Material, Department of Food Science and Technology prepared by Sri S. Chenna Kesava Reddy, part-time teacher, CFST, Pulivendula and Mrs. R. Preethi Sagar, part-time teacher, CFST, Bapatla.
4. Acharya NG Ranga Agricultural University College of Food Science & Technology Pulivendula -516 390
5. Diet, nutrition and the prevention of chronic diseases. Report of a Joint FAO/WHO Expert Consultation. 2003

6. What is serving? Food and agriculture organization of the united nation. Last accessed on 2010 Dec. Available from: http://www.fao.org/english/newsroom/focus/2003/fruitveg2.htm.
7. National Institute of Nutrition (2011) Dietary guidelines for Indians: A manual. Hyderabad
8. Annual Report (2011), Ministry of Agriculture, Government of India, New Delhi
9. Fruit and Vegetable Preservation by N. P. Singh, Oxford Book Company, Jaipur, India ISBN 978-81-89473-12-9, First Published 2007, Oxford Book Company 267, 10-B-Scheme, Opp. Narayan Niwas Gopalpura By Pass Road,Jaipur-302018

Appendix

Processing Equipments, Utensils, Preservatives and Packaging Materials

The success of processing and preservation is based on skills and quality of equipments used for preparation of different fruits and vegetable products. There are several equipments available in the kitchen and those which are not available they can be made available.

Equipments and Utensils

1. Stainless steel knife
2. Stainless steel fork
3. Stainless steel spoon
4. Wooden spoon
5. Stainless steel bowls
6. Small wooden board
7. Stainless steel filter
8. Wooden Ladles
9. Graters
10. Slicer
11. Juice extractor
12. Measuring cylinder
13. Hand refractometer
14. Muslin cloth
15. Balance
18. Stainless steel cooking pots-different sizes
19. Gas cylinder with Gas stove
20. Beaker
21. Brush
22. Table watch
23. Thermometer

24. Pestal and Marter
25. Stainless steel coring knife
26. Stainless steel peeling knife
27. Working tables
28. Stainless steel household sieves
29. Aluminum pots
30. Hand-operated pulp extractor
31. Wooden lemon-squeezer
32. Stainless steel skimmer
33. Sun dryer (tent type)
34. Sun dryer (cabinet type)
35. Standard wood sun drying trays
36. Bottles
37. Crown tops Jars
38. Screw-tops for jars
39. Wood (or other available fuel) heating plates
40. Hand-operated capping device (capper)
41. Work bench
42. Stainless steel vegetable cutter
43. Glass jars with lids
44. Jam bottles
45. Sauce bottles
46. Bowls with covers.
47. Wide mouth jars
48. Stainless steel Jug
49. Glass jars, various sizes and screw-on caps
50. Solar dryer

Preservatives

1. Food grade Sodium benzoate
2. Food grade Potassium metabisulphite
3. Food grade Sorbic acid
4. Food grade Citric acid
5. Food grade Acetic acid
6. Sugar

7. Common salt
8. Food grade vinegar
9. Mustard oil

Salts for Maintain Natural Colour

1. Sulphur
2. Magnesium oxide
3. Sodium carbonate
4. Potassium metabisulphite (KMS)

Common Spices

1. Cumin
2. Fennel
3. Nigella
4. Asafoetida
5. Black pepper
6. Cloves
7. Garlic
8. Onion
9. Chilli
10. Cinnamon
11. Small cardamom
12. Large cardamom
13. Turmeric powder
14. Ginger powder
15. Fenugreek or Methi